THE MAN, THE MACHINES, AND THE LIFE

A TOOL AND DIE MAKER'S STORY

By

Seymour Lehrer

Summerland Publishing
Offices in Salt Lake City, UT and Summerland, CA
www.summerlandpublishing.com
Email: summerlandpublishing@gmail.com

Printed in the United States of America.

ISBN #: 978-1-7375730-7-4

Library of Congress # 2022903829

TABLE OF CONTENTS

Preface

I realize the title of this book may not be what many might consider 'sexy', and perhaps the majority of people would throw this book away. But I urge you to stay with me for at least the first page, and perhaps I can show you just how vital the tool and die industry is to our everyday existence.

Also, I would like to dedicate this book to my wife Shirley, my two kids, Karen and Ellen and their families, who are the most important people in my life. I say this, of course, because of the fact that they have been with me every inch of this adventure; but I also say this to stay out of trouble with Shirley…

PART I

CHAPTER ONE: How the Tool and Die Making Industry Engineers our Modern World

Why in the world would anyone want to know about the Tool and Die making industry you might ask?

Little is commonly understood as to just how central this industry is to the comforts and conveniences of our daily life! Let me explain…

On any given day in our modern world, a member of the family comes home from work, takes the key, and puts it into the front door lock, hears the click, presses the little lever and the door swings open. All parts of the lock as well as the key were stamped on dies in small punch presses. The door itself is mounted on two heavy gauge steel hinges, which were also stamped in dies.

The individual reaches over and touches a light switch in the wall, and a light magically turns on. All the parts of the light switch were stamped on small dies. The socket that the bulb is screwed into was also made by stamping. They then move over to a thermostat mounted on the wall and adjust the temperature gauge to provide for ideal comfort. The thermostat is connected to the heating and cooling unit which is mounted in the ceiling. Most of the parts of this entire system were stamped out of sheet metal.

He or she walks into the kitchen. Mounted in the counter is a large white painted sink with a handsome looking valve set above it. If you push the valve one way, hot water comes out from a hot water heater in the attic. If you push the valve the other way, cold and perhaps filtered water comes out. Again, all these parts were made by metal stamping. Additionally, the bottom of the sink has a large pipe and garbage disposal machine with an electric motor and many smaller stampings within it to grind and dispose of the waste.

On the counters, there are many kitchen appliances, such as a coffee maker, a toaster, a blender, and a range top stove among other items. Some of the parts for these are made in plastic molds. The metal parts are stamped on dies. The individual then opens a drawer that contains all kinds of silverware, all of which was stamped in dies and were then polished and plated. The same is so with knives - the blades themselves were also stamped.

Now we open up the refrigerator, the body of which was stamped on huge dies. At the top of the refrigerator is the ice cube maker and the tray of loose ice cubes, all made from metal stamping. The reader may be interested to know, that General Electric spent over 50 years to develop the technology of the ice cube maker, a convenience that we so casually enjoy today.

Basically, everything I have described thus far was manufactured with tools and dies that supply the greater household industry.

Elsewhere in the household, one can find pins and needles that were made in tiny dies. Also, one will likely find paper clips, thumbtacks, staples, and staplers, also made in small dies.

Perhaps in the tool shed will be found screwdrivers, pliers, wires, hammers, wrenches, tacks and nails, screws, and a small assortment of fasteners. All these items were manufactured in dies of varying sizes. Additionally, the house itself was constructed by parts that were also stamped.

In this first section, I have tried to illustrate how vital the die making and stamping industry is to the existence of the modern household, that includes conveniences which many take for granted today. In fact, this crucial industry is central to the entire construction industry itself.

CHAPTER TWO: My Beginnings

In 1941 when I was 14, my brother Dan, who was 15 years older than me, hired me to come down after school to sweep the floor in the machine shop that was owned by Frank Pachmayr. Dan brought a lot of Douglas Aircraft work to Frank, who had hired him to be the shop superintendent. I got $.15 an hour. A month later, I got a dime raise. They started to teach me how to run machines. I loved the work. My wages went up to $.60 an hour. On summer vacations I worked 10 hours a day, six days a week. I saved up a couple hundred bucks in the bank. This kept up till the end of the War in 1945. I was 18. I got jobs in several other shops and gradually became a first-class machinist.

A couple years later I met Shirley. We were sitting on a pile of hay in the back of a truck on the way to Ocean Park. It was called "The Hayride". She was sitting next to me, and I started gabbing with her. She was very quiet. I was in love. By the end of the day, Shirley invited me to come see her on a Sunday. She lived on a chicken farm in Reseda. I came to see her once a week, but it was a long schlep from Boyle Heights—35 miles. So, I got a job at Lockheed, (which was halfway between Boyle Heights and Reseda) starting at $1.35 per hour.

A month later, I got a $.10 raise, increasing my status to a Class 2 operator. Three months later I received another dime raise.

When I told Shirley that, she went out and talked to her mother and came back and told me, "Well, maybe once a week, instead of driving home, you could stay here and drive to work from here."

I went ahead and did that, except I did it on Fridays, so that I could stay over on Saturday, which her mother agreed to.

This went on for a year, and one day I took her out for dinner at a very fancy place. We both got lobster, which was the first time either one of us had had it. While we were eating the lobster, I leaned over and told her that I thought we should get married. She quietly asked me if I had an engagement ring.

I said, "What's that?"

She said, "It's a big diamond and it announces that we're going to be married."

I told her that we would go see Mr. Glucoft. He was the father of one of Dan's friends, whom I had previously met, and he was a jeweler in the Jewelry Mart in downtown LA. We went there the next day and I told him the story about the engagement ring, and that we wanted one. He looked at her hand and then at her face and he asked me how much money I had. I told him I had $75. He thought for a minute and fumbled around in the showcase and took out a white gold band that had a row of five 5-point stones. He put it on Shirley's finger, who almost passed out. She thought it was the most beautiful thing she'd ever seen. She loved it. I paid $75. I was a hero to Shirley. We were married June 11, 1949. A month later, Shirley turned 20 and three months later I was 22.

During this period, I got close to my supervisor, Joe Miller, and we had the opportunity to talk.

One fine day, he said, "Seymour, I looked up on your employment record and found that you worked for Frank Pachmayr for over three years. That was when I discovered that you knew just about everything needed to be a machinist at Lockheed. But you have to understand that this is a closed shop, and the Union makes it difficult to put an advanced man in the right position. And your problem is you're just too damn young and haven't been working here long enough to get to be an "A" machinist. Seymour, I'm going to give you some advice. Get the hell out of Lockheed and get yourself a job at a tool and die shop. And if you breathe a word of this conversation to anybody and I find out, I'm going to kick you in the balls."

Six months later I was involved in a 500-person layoff at Lockheed, and I got a job at Standard Tool & Die. In January 1950, on the first day of my work at Standard Tool & Die in Los Angeles, I was running a big milling machine and a young guy walked over to me with a smile on his face and said,

"Hi, I'm Del Molinari. What's your name?"

I said, "Seymour Lehrer".

With a question mark on his face, he said, "*Seymour?* I've never heard of that name." Del said. "Fred, our foreman, told me that you worked at Lockheed."

I told Del that I worked there for over two years and got canned with 499 other people in a big layoff, and that is the reason I am here.

Del said, "I'm a diemaker. Freddy said you're a machinist."

And I told him that I was. That started a lifelong, close friendship. Over the next few months, we learned all about each other. And we started working together on weekends doing odd jobs.

Our Very First Job, A Rooftop Vent Cap

First, I worked with Del in one of his friend's shops. And later he worked with me in my brother Dan's shop. This continued for a couple of years and in June of 1953, Del and I got a $2,000 job to make the dies for a household vent cap, which is the cap that is used to divert rain away from the pipes of the house. We were working on this job in Dan's shop. One day, Dan came out and said he wanted to talk to us. He told us that he had decided to sell his shop. He was not making any money out of it, so he was offering it to us. We asked him how much he wanted, and he told us $10,000. That stopped us. The next day we told Dan that we wanted to rent the shop from him, and we were willing to pay him $80 a week net and Dan should pay all expenses. We told him we would pay him $160 in advance for two weeks work. Dan agreed.

We each had about $1,000 saved up for business purposes. And we quit our jobs at Standard Tool & Die. We named our new business 'Commercial Tool & Die.' This happened in 1953. It took us three months to collect our money from the $2,000 job, during which time we were working like hell. We got several other jobs—a couple of which we did not make any money on. We each took $50 a week home from Commercial Tool & Die to live on, which was less than half of what we were making at Standard.

One of Our First Jobs, A Special Sawing Machine to Cut the Box Spring of a Mattress

On December 8, 1953, Shirley and I were involved in a large auto accident. Shirley was driving and I was sleeping in the front seat, and our two kids were both sleeping in the back seat. A drunk backed a huge truck right out into a line of traffic. We ran into him, as well as two other cars. The driver of one of the other cars was killed and the third car smashed into the back of our car. I went through the windshield. My knee bent through the dashboard. Our car was totaled. We wound up in LA County Hospital. I was off work for more than two weeks. When I returned to work, we were informed that Dan had not paid his taxes and the IRS was going to close his business. So, we had to get all our stuff out of Dan's shop, including all our work in process, which we moved to the garage at my house.

In the next two weeks, we rented a 2,600 square foot section of a larger building. We had an ace in the hole in the form of Dave Rudin, an electrician and good friend of mine, who was Shirley's uncle. We explained to Dave what our problems were, and he told us that he could knock this out for us in a couple of weeks. He wired up the place, hung lights, and within a month we had purchased over $20,000 worth of machinery, paying 20% down and financing the balance over three years. By the end of February 1954, we were in business in our own shop. We owed the bank $850 per month.

Our First Shop, Where It All Started

During the next month after the wreck, I had had to spend a lot of time in and out of the hospital, UCLA Stein Eye Clinic, where they had to take glass out of my eye. I am grateful to them for saving my eye. Three or four days out of the week, I was laying in bed. My shoulder had almost gone through the door, and I could hardly move it. To this day, in fact, I have very bad arthritis in that shoulder. Also, my doctor had to re-open my knee and reconstruct it, as the emergency doctors had not properly sewn my knee the first time right after the crash. That knocked me out for a good week!

So there I was debilitated, and yet we had just moved into our first building. The clock was ticking, the money was owed to the bank, and we had to make it work! But I could not help run the machines and produce the product. Del said to me, "Seymour, why don't you be the manager and run the company and I'll run the shop."

I agreed. Still, however, it was hard for me to get around. The first couple of months, I was on crutches. Just for me to walk I was in a good deal of pain. But I couldn't think about that, I had to make this company work; Del was depending on me. So, I put my mind to it. I was at work by 6 a.m. and I was constantly on the go purchasing materials, tools, and machinery for the company. Plus, Del and I had to go out and prospect for new customers. I didn't finish each workday until 7 at night. I had to put the pain in the back of my mind.

That is how we got our start, but it was rough going at first.

I still have my income tax reports from that era. In 1952 and half of 1953, I showed good income from working at Standard Tool and Die. But in 1954 we made next to nothing. We were master machinists and craftsmen, but we were not businessmen.

So how did I live you ask?

I had sued the owner of the big truck that ran into us. I got a settlement check of $8000. I wrote a check to the company for $4000 and used the balance to buy a house on level ground at 266 Twickenham in Monterey Park. I sold my old house and I got $4000 for it. That's what we lived on.

Little did we know though, that our next move was going to lead us to the help we needed to actually make a successful business.

Let me explain…

In December of 1954, Commercial Tool & Die had joined the Southern California Tool & Die Association (SCTDA). At the first meeting we walked in, Bob McCullough rushed over and grabbed me in a bear hug and said to the assembled members:

"I want to introduce you to Seymour Lehrer and Del Molanari, his partner. Seymour worked for me when I was the foreman at

Pachmayr. He was hired to sweep the floor after school. He did such a good job that a month later I gave him a dime raise."

I interrupted Bob and said, "I thanked Bob for almost doubling my wages."

Everybody started laughing at the thought that a dime doubled the wages. That broke the ice. Bob continued to say that he did not mind telling everybody that he taught me every goddamn thing that I know about machine work.

I told him that yes, "you did teach me a lot, but you didn't tell me a goddamn thing about being in business." Bob said that he never even dreamed that *he* would ever wind up being in business, let alone that I would ever be in business, so forget that!

This introduction turned out to be the fortuitous break we needed to really make a go of it in our industry. It was the connections that we began to make here that led us to our education. But I will explain more about that in chapter four titled, "Getting Educated in Business."

CHAPTER THREE: What and Who Is A Tool & Die Maker?

A tool maker is an accomplished machinist who can assemble contraptions called jigs, fixtures, and special machines that are designed to assemble specific parts. Any object that needs to be machined and assembled into a product requires the particular tools that the tool maker can make.

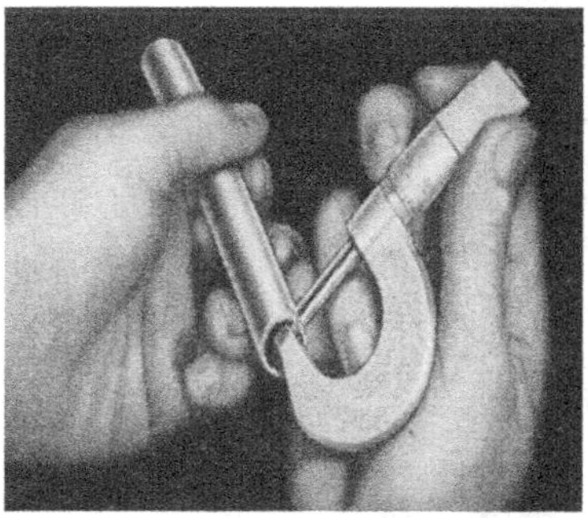

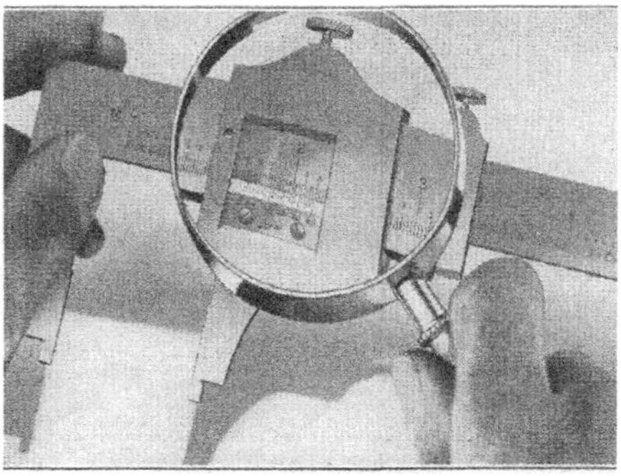

The die maker needs everything that the tool maker has plus several more years training in the making of the punch press dies. Most sheet metal objects are made on punch press dies. The punch press die is an object that goes perfectly up and down on a couple of guide pins so that the punch, at the top, enters the die exactly in the right place.

Parts of a Drawn Die

One of Our Early Punch Press Dies

Vertical Mill 1

Vertical Mill 2

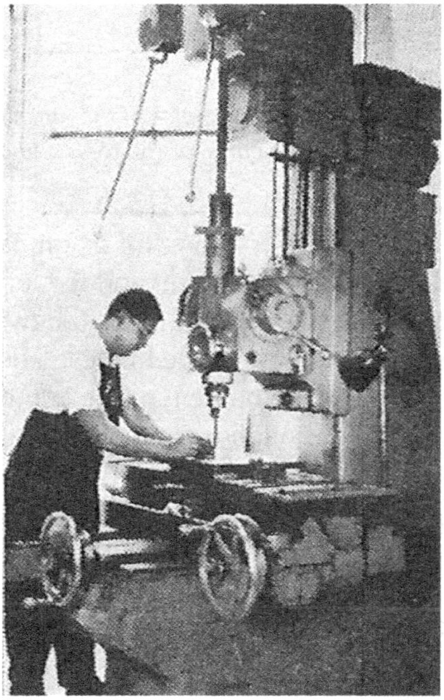

Jig Borer

36 Inch Blanchard Grinder

20 Inch Engine Lathe

The above pictures are all examples of the machine tools required to make the parts for the tools and dies.

To manufacture a door hinge, for example, a punch press die is required that will stamp both the right and the left-hand parts completely. From there, the tool will take the two parts and automatically assemble the pin in the hinges. Historically speaking, a great deal went into the development of the common door hinge. Initially they had to figure out how to make a pair of hinges and automate the process. Then, they had to figure out how to automate the packaging of a pair of hinges into a cardboard box that could then be sold to the construction industry. The evolution of this process took about 15 years and was accomplished roughly around 1910. The hardware industry was developed in the Northeast states and is highly specialized.

Over the years many objects have been developed that are manufactured by automation without ever being touched by human hands. As an example— the electric light bulb, designed by Thomas

Edison was too delicate to be produced manually, and therefore had to be manufactured by machine. It took eight years to perfect that process. Even the hypodermic needle is manufactured by automation. These objects are the culmination of the tool maker's art.

Industry had developed the tool engineer and the manufacturing engineer—both of whom in the advanced degree are trained in a couple of northeastern colleges. The tool engineer is a lazy tool maker who sits in a clean office in front of a design board and constructs the advanced procedures required by this sophisticated industry.

The extreme example of mass production is the modern automobile. The engine block of a car is installed into a gigantic machine that automatically clamps it, runs the various areas of the block through milling machines, drills and taps it, bores the cylinders, finishes the cylinders, and ejects the block into another gigantic machine that assembles the entire engine. All the various parts are machined and most of them assembled onto the block. This gigantic machine is about 300 feet long and can produce an assembled engine every fifteen minutes.

The body of the car is stamped on a selection of presses from small to huge--the largest being over 2,000 tons stamping capacity. Each major part of the body must go through several operations. The final assembly line can be over a half a mile long, assembling the entire car together also in fifteen minutes.

In contrast, take for example the common stapler. Most people don't really understand what all went into perfecting this useful little tool. Staples are glued together and stamped several hundred at a time and assembled in a box. The row is slid into the stapler which slices off one staple at a time. The stapler itself is made by the die maker in a special machine that stamps out the parts and then assembles them together. The staples themselves are made by the tool maker. The expansion of the paper stapler became much larger staple type machines that could staple two boards together, again another vital tool that modern construction depends upon.

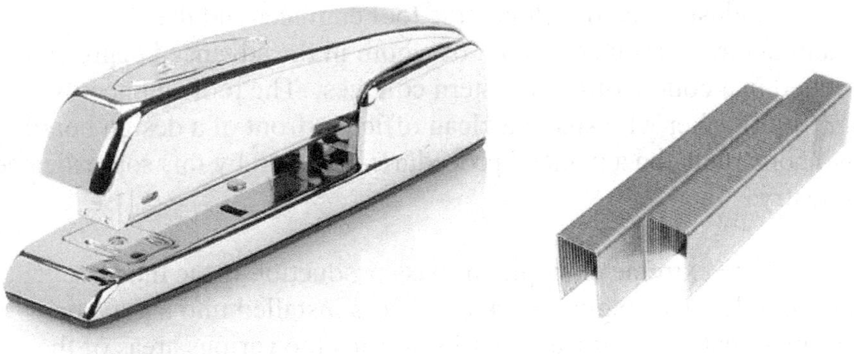

As you can see, everything from little staples on up to large vehicles and construction items depend solely on what the tool and die makers do. None of any of this technology would even be possible without the tool and die maker.

CHAPTER FOUR: Getting Educated in Business

There are many "do's and don'ts" that you must remember in order to survive in business. As I mentioned before in chapter two, we had joined the Southern California Tool & Die Association. (SCTDA) The great advantage that we rapidly realized we had was being able to listen to and learn from our fellow association members. Often, somebody would drop a pearl, and the next day Del and I would discuss the subject at length. This went on at almost every meeting, because we realized that all these guys had gone through the ups and downs of the business, and we had a lot to learn from them.

About three years after we started our company, we were sitting in the little office having lunch.

I told Del, "Well, congratulations. We just got done paying off all our machinery. And that's a real relief." Though I was happy to announce that news, I still had a slight look of worry upon my face.

Del then asked, "Seymour, what's troubling you?"

I said, "My salary is not sufficient and I'm having a tough time getting by financially. Every month, when my $40 mortgage payment comes up, I go into a sweat." Del reached in his pocket, pulled out a pile of bills and laid down five twenties. So, I asked him, "What the hell is that?" He said he was giving me $100 to tide me over. I told him I'd pay him back.

The next day Del sat down and said, "I've given a lot of thought to the conversation we had yesterday. Do you remember after you got your $8,000 settlement from the car wreck case?"

I said, "I sure do."

He said, "You wrote a check for $4,000 when the company was dying for money. You then took the remainder and bought your little house on level ground, because you couldn't make it up the steps in your old house." I told him that I remembered all that. Del said, "That really bothered me a lot."

And I told him, "Did you forget that a couple of years later you took out a $2,000 second mortgage on your house and gave it to the company when we were in a bind?"

Del said, "That's right, but I still owe you $2,000 and I'm going to pay you back at the rate of $50 a week. So here is $200 more than I gave you yesterday. That takes care of six months payments that I'm going to make to you." Sure enough, he kept his promise for two years or more. I thought about what a guy he was that I picked for a partner.

One day I asked Del, "How are you managing to come up with all this extra dough?"

And Del said, "Pat. You know that she always made more money than I did. She has a private bank account, and I have no idea what's in it. But Pat and I have a joint bank account that we both put money into. But she puts in three times as much as I do. That's where I get my money."

The years between 1953 and 1960 were spent on an awfully expensive education. Like I mentioned in chapter two, we were very accomplished craftsmen. We were NOT businessmen. We made all kinds of mistakes, but we tried to never make the same mistake twice. We had already joined the National Tool & Die Association and had made friends with some of the old timers. One of those individuals was Carl Weitzel. Knowing that we needed help on the business side of things, I called Carl up one day to see if he could help us. It was

1957 by this time. After looking over our financial statements, Carl threw up his hands and said, "you need more help than I can give you!"

He made a phone call to Harold Murdock, the President of Arrowsmith Plastic Tool. He was known as one of the smartest businessmen and accountants in the industry. Carl brought me to Harold's office. Harold spent a full day with me and taught me more in one day than I had learned in 7 years.

Harold told me that the worst problem was that we were lending credit to the wrong customers, and we simply were not getting paid on time or not at all. Harold said, "the best friend you might have in business is the credit manager of your bank. What you need to do is before you do business with anyone you need to get the approval of your credit manager. You have to realize that you are his client, and he does not want you to lose money. The second thing you have to realize is that you need to join the credit managers association. If you did, you would know everything about everyone you were soliciting."

Harold went on to say, "I do not do business with anyone that my credit managers will not approve, and we never lose any money." That was the biggest thing I had to learn. I would call the credit managers up and they would tell us exactly who we could successfully give credit to. Fortunately, there were plenty of customers out there who did have good credit who took the place of our deadbeat customers.

Harold also told me that I needed to charge at least $15 an hour to stay in business.

With that knowledge under my belt, I became more selective, and we finally started making good money.

I mentioned in chapter two that I still have my old tax returns from this era. In 1952 I cleared $6,242 from working at Standard Tool and Die. In 1953 I had made half that from working only a half year at Standard. 1954 was a bust, with only $146 income. In 1955, I actually made $5,337, a big jump but still less than what I was making working at Standard. I kept inching forward as best I could, but it was not until 1958 that I started to show the fruits of our education in my income.

Building our customer relations

The formation of several of our customer relationships was also a key to our success. A few of those came about by purely happenchance means such as our work with Travis Electric. Let me illustrate.

One fine day a truck drove up to our door and I went outside, and I saw a pile of steel that was going to Travis Electric. The driver said, "not yours this is going to Travis." The next day the same thing happened. He unloaded a couple of bars of steel, and I noticed the big pile that was going to Travis Electric. I called my bank and I asked them, "who is Travis Electric"

And my banker said, "Are you joking?"

"No, I am not joking," I replied.

He went on, "Travis has been one of our big customers for over 40 years; they are a recommended customer."

I phoned Travis Electric and found that they were two miles away from our shop. I went down there and asked to see the tool engineer. The lady told us, "We do not have a tool engineer; we have a foreman of our shop."

"Ok I'd like to talk to him," I said.

An old guy came out and I introduced myself to him. He said, "Come on out I'll show you my shop."

His shop had older machinery than even our shop had. And he had about 10 guys working for him. But to me it didn't look like they were working. He asked me what I would charge an hour to do repair work for him. That put me on the spot. I lied to him, and I told him we'd work for $12 an hour.

He said, "My cost is $14 an hour and I'll pay you that."

He was showing me work that had to be done and right off the bat we loaded my truck with two big dies. He said, "Take them back to your shop and I'll meet you there", which he did the next day.

We discussed the job and he said, "It's impossible to depend on repair work of this type until you take the entire die apart." He said, "Take it apart and I'll come to see you tomorrow."

Next day he showed up at our shop and he looked like he was drunk. He asked me, "Do you by chance have a bottle of whiskey?" Apparently whenever he had one or two shots of whiskey he could run for hours.

We took the die apart and discussed what it would cost to rebuild it. He said, "Get started, we need it in a hurry." I put our three best die makers on the job, and I billed him at $14 an hour plus just a little extra. We shipped the die to him, and he called me up and he said, "I like your work, and I have a lot more for you."

Travis Electric was an old line of industrial electrical equipment, it was a very successful old company with over 100 employees. This started a relationship which actually lasted for more than five years. Travis had employed a die design company to design a set of dies for a totally new product, which was designed for individual small houses that were being built by the thousands. And they had circuit breakers in them. The foreman came to our shop with a roll of blueprints which we laid out and started studying. Del and I decided we understood what the construction was to amount to and tried to estimate what the cost would be. We discussed it with the foreman, and he said, "You two guys are full of shit, you can't build that die for what you're talking about without going busted!"

I said, "We've done a lot of work for you, and you trust us and I trust you, so help us out on a price."

He said, "Put 20% on top of your price and I'll get the job."

The job amounted to $30,000 and we made money on it. We finished it up and brought it down to Travis. They set it up in the press and started production. What happened was after a month, he came up to the shop. I gave him a couple of shots of whiskey, and he told me, "build a second die." We told him we thought we could refine the die and make it better than the first one. He said, "go ahead and build it and refine it."

A month later he called me up and said, "How you doing on the new die?"

I said, "we're almost done."

He said, "Put everybody in the shop on the die, it has to be finished immediately."

A week later he called me up and asked, "How you coming? We need the die!"

I said, "We're almost done."

He said, "Finish the goddamn thing and ship it!"

Which we did. What had happened is they had wiped the first die off almost ruining it and had to have another die made. They sent the old die back to us and I put everybody in the shop on re-building the first die. After two or three weeks, he called me up and he said, "I hate to tell you this, but my guys have just wiped the new die out." Dies can be very complicated, and a little thing can make it go off kilter and ruin it. So, we finished the second die and rebuilt the first die again.

We did a lot of work for Travis and made very good money. In fact, Travis was one of our breakthrough customers who really turned our business and income around.

Another company also helped us turn a new corner in our business, Detroit Stamping and Manufacturing Company. One of the credit managers whom I became acquainted with suggested that I go to see them as they were only a few miles away from us.

I did and was introduced to the Tool Room Superintendent. Detroit Stamping Company was a wholly owned division of Ford Motor Company. The plant had over 500 people in it and manufactured all the metal parts of the seats for the Ford Motor Company cars that were made on the coast. Making a long story short, we did an enormous amount of work for Detroit at $15 an hour, and basically between Travis and Detroit, it made our company.

One fine day in 1960, Del came into my office, and I told him I have a problem and he said, "What's your problem?"

"Over the last seven years I borrowed $22,000 from my wife's parents and $3,000 from your parents, and $4,000 from my parents. We now have $70,000 in cash and I want to pay back those loans that had been made to me and you."

Del said, "the loans were made to you, but they were really made to the company. Take $35-$40,000 and pay it back."

I said, "Wait a minute. Do you realize what you're talking about?"

He said, "Yes I do, just do it."

Seymour's Father / 1st Week in Business
Notice the black walls

I thought, boy what a guy! And I paid them back. And we paid off the bank for the second time for the machinery.

By 1960, with all our problems, our education was near complete. We had finally paid off the debt for all our machinery, and we had some money in the bank. That was when our business took on a whole new level; let me explain.

Del with Bench Master Half Press which he made 1953

One day Carl Weitzel visited our shop and made a surprising suggestion. "Seymour, I'm going to give you some advice. In my opinion, I think you guys are too good at your work to be a small-job tool and die shop. What you need to do is become a manufacturer of your product, and to go into the stamping and manufacturing business." That was a mouthful. Del and I started talking about what our friend said. Easier said than done, but it stuck with us. Four years of saving money, and many preparations later, we did finally graduate into having our own stamping and manufacturing business.

Seymour & 75 Ton Bliss Press 1960

CHAPTER FIVE: Tool & Die Apprenticeship and Training LA Style

While we were forming and establishing our manufacturing business, I also became more heavily involved in the associations that helped CTD oversee our industry. After we had joined the SCTDA, I became a member of the Joint Apprenticeship Committee which was running the Apprenticeship Program for Southern California. Joint meant that half the members were from the Machinist Union, IAM (International Association of Machinists), and the other half were from members of the SCTDA. There was a tremendous conflict between the Union members and the SCTDA members due to the extreme regulations enforced by the Union. Our three guys hated the guts of their three guys, and vice versa. Aside from the six, there was one consultant from the State Department of Employment and one from the LA City School system, which ran the nighttime instruction program. After several meetings, the SCTDA members announced that they were having a lot of trouble with the Union guys and were going to quit if nothing was changed.

The Electing Members of the Board of the Apprenticeship Committee

Somehow, I got along with both the Union members and the SCTDA members. Because of this, all eight members elected me to be chairman. Right off the bat I understood the problems, and I did my best to resolve them and keep the organization together. I held this position for a few years, during which time Carl Weitzel had reported my activities to the National Apprenticeship & Training Committee (NATC), of which he was a member.

In 1956 Commercial Tool & Die joined the National Tool, Die & Precision Machining Association, NTDPMA--a large industrial organization with over 300 members from all over the United States and several in Canada. NTDPMA was based in Washington DC and operated the NATC. Both had been in existence for over sixty years, and still are today. NATC had printed numerous books concerning the instruction of apprentices. They were the keystone of that type of training in the United States.

The local members that were also members of NTDPMA formed the LA Chapter. In late 1962, I received a letter from NATC advising me that my efforts as the Chairman of the Joint Apprenticeship Committee were being recognized, and that they suggested that I replace Carl Weitzel who was the representative at that time of the LA Chapter.

Carl and I had become close friends and he was now 82 years old. He called to tell me that he was the one who had wanted me to replace him, as he was going to resign at the next meeting, where he wanted to introduce me. I then took his place in the beginning of 1963. By that time Commercial Tool & Die had come way out of our poverty-stricken period and we were personally able to get extra money or more salary to do whatever we wanted—like travel to Washington DC. At the first meeting where I was introduced and Carl resigned, we were sitting around a large 20-person table, and in front of each person was a card with the name of the individual and the company he represented.

I was sitting next to Carl looking across the table and I said, "Pratt & Whitney, Brown & Sharp, and Moore. My God, how could I be sitting across from those guys?"

Carl said, "Those guys are from the original companies that started out the same as you did."

And I said, "Pratt & Whitney, who makes airplane engines?"

He said, "Yes, that's them and they started out the same way you did making machine tools." I was amazed and overwhelmed!

The Committee met every six months and in between there were communications sent out keeping the members apprised of what was going on in the industry and in the apprenticeship in particular. Ed Hardman worked for Nationals, as the Director for Apprenticeship and Training and was appointed as head of National's efforts in that regard. He used to meet with the US Department of Labor people discussing the problems of Industry as it related to labor. One of the problems was that President Kennedy's efforts to put a man on the moon and bring him back alive had caused a very definite shortage of skilled labor. The top people had been hired by the huge companies that were working on Kennedy's plan, and the members of Nationals were complaining to labor that there could be a national emergency due to the fact that there was a shortage of skilled workers.

Lyndon Johnson (the Man of the People) as a representative, and then a Senator had met with the Department of Labor repeatedly. Labor had come up with a plan, which was called the Manpower Development and Training Act (MDTA) of 1963. After Kennedy's assassination, Johnson became President and immediately his people reviewed the MDTA program. He almost immediately had Congress approve the bill, mainly because there was almost no money involved. The program said we will administer training programs, we will pay for instruction and maintenance, but we will not lay out a dollar for bricks and mortar nor furnishing facilities. These must be provided by private industry. Congress loved the program. The details of the program were outlined to the members of the Committee. The problem was that we didn't know how to get started, and actually for about a year, almost nothing happened.

In 1965 after we moved out of Jesse Street and into 51st Street, Del and I bought a 3,500 square foot garbage two-story building a few blocks away from Jesse Street. We had fixed up the building and I called Ed Hardman, who was the training coordinator from National, and explained to him my plan, which was to turn the little building into a training center. There were many problems that had to be solved. Ed told me he was going to come out to see the building and listen to my plan, and he did. My plan was to hire unskilled workers, train them, put them through school for maybe six months, and then get them jobs in members' shops. Ed came from DC, spent a few days in LA during which we had several talks, and he went back to DC.

He got back to me on the phone and said, "Labor loves your plan, and I am positive that they will support you. I have several things that you will have to do to improve the building to their requirements, but it does not amount to much. But I want to tell you that the key thing is that the new construction across the street that the city was building as a housing project to house primarily poor families, is where we can get our pool of candidates."

I told him, "Hold on. Did you forget that I wanted to get guys that had graduated from LA City schools and had taken shop courses, which was recommended by our consultant from LA City school system?"

Ed told me that he'd forgotten all that, but that can only help it to work. So, we ended up doing a combination of the two. We were subsequently able to get 3,000 people jobs over the years as a result of this program.

By this time, the local members that were interested in apprenticeship had determined that we had to leave the union and start our own apprenticeship program. I met with the State consultant and the consultant from the LA City School system and explained the problem. They got together and a few days later they asked me if the local SCTDA would sponsor a program which would not include IAM (International Association of Machinists) shops. I told them that it was not a problem because there was not one tool and die shop in the entire area that was Union. They said that the present joint committee was going to remain in existence. I had a meeting with the President and the Vice President of the local organization, and they had informed me that at the next meeting it would probably be put into action, and they would apprise everyone about the subject. Sure enough, at the next local meeting, I was asked to get up and lay the whole plan out. Instantly people started yelling to call for a vote. The President called for a vote, and it was absolutely unanimously adopted, and our private apprenticeship commenced. I was also elected Chairman.

We had one additional event that especially motivated us to leave the Joint Committee. We had been developing a shop with the joint members that was set up at the Pan Pacific Auditorium. This had been set up and worked on for almost a year. Several members of the local organization had worked like hell preparing everything that had to go on and many companies that sold machinery were going to donate several machines. We were going to get them moved, set up and run a contest for the apprentices so that we would advertise what we were doing. This took an enormous amount of work and was a fantastic success. People by the thousands attended the contest and crafts show. The problem was with the Union. They were supposed to participate in the contest and not one Union guy showed up. That really pissed off our members and that was one of the big reasons we finally split.

Back to my plan at the little building… Ed Hardman told me that there was no question that we would definitely get a contract from the Department of Labor, but the contract would go to National, and as the local chapter of National, this would be our baby, but also sponsored by National. I had discussed my plan at a Board meeting of the Local. The Board decided that they should consult lawyers to determine how the program would work. The lawyers had determined that we should establish the LA Training Center Trust (TCT),

Three Officers of the Tool & Die Association
Seymour Lehrer, Bob Knudson, Laurie Weitzel (Carl's Nephew)

The Winner of the Apprenticeship Contest, Chuck Meisner Who Worked for Commercial Tool & Die

sponsored by the local group. The TCT would be responsible, and no individual company or person could be sued for any reason. At the next meeting I got up and I explained everything that was being considered by the TCT for the building. I explained that the Association members would have to come up with the equipment to furnish the machine shop. The Main Shop area of the Training Center contained all the basic equipment necessary for well-rounded instruction in all facets of machine shop practice, including surface plate inspection techniques. The class size was usually 20 young men and equal opportunity was offered to all.

Within two weeks, several of our members, including Commercial Tool & Die had donated almost enough equipment, and a couple of our members had contacted machine tool suppliers to suggest that they should also donate machinery. And that happened. Ed Hardman had arranged through the LA City School system to have Mayor Yorty cut the ribbon at the inauguration of the school in 1965.

The mayor's office had talked to the police about the parking problems on both sides of the street. They had to work out a plan that allowed us to park from 9:30 am to 3:30 pm, which was perfect for us. The City School administrators had worked out aptitude testing and National had all the instruction material prepared for years. The school then went on successfully for many years and hundreds of students graduated and went on to good paying careers.

Seymour Accepting Award for Service on the Apprenticeship Committee

CHAPTER SIX: The Growth And Our Big Break

Alongside the development of our training school, our own company experienced tremendous growth beginning in 1960.

One day Del sat down and said he had something to tell me. And that was that he and Pat were going to get a divorce. I sat there stunned. I couldn't imagine such a thing. I asked him how could that have happened?

And Del said, "For the last six months, we had not been sleeping together. And the problem is that she requires much less sex than I do. She told me that she wanted me to support the kids to a degree, but not everything. But she wanted me out of the house. She also told me that if I give her a hard time, she is going to take the kids back to Riverside, where her parents live in a big, old farmhouse, and they would love to have her and the kids for themselves." He told me, "She's got me by the balls, and she knows it."

I asked him what was going to happen to the business and what was going to happen to the Jesse Street building we just bought. Del assured me that Pat had no interest whatsoever in either the company or the real estate. Del rented a room in Huntington Park and moved out of their house.

1960 was a good year. We made money and kept it. In 1961, 1962 and 1963 we did over $500,000 of business per year and made big money. We had $200,000 saved up in the bank and business was great.

We had accumulated a pretty impressive list of good customers. Several of them were manufacturers of items that went into the construction industry, such as air-conditioners and heaters, water

Seymour with a 4 Headed Router Made for General Motors

Del with an Inspection Fixture
Made for American Electric

heaters, garbage disposals, electrical equipment, lighting fixtures, outdoor furniture, and steel fabricators. We worked with many companies including Kaiser Steel, Norse Thermador, Marvin Electric, Fletcher Aviation, Day & Night Electric Appliance, Travis Electrical Equipment, and Kaiser Aluminum, to name a few.

In 1964 the miracle happened. The foreman of Detroit Stamping came to us and to give us a job manufacturing the dies used for the stamping of a new line of mop presses. He was doing the manufacturing of the mop presses for a company called Continental Janitor Supply. We started on the set and completed them in about a month and a half.

One day, the foreman of Detroit called me again and told me he had lost the job of making the mop presses for Continental. I called up Continental and asked them what they were going to do with the dies that we had made for them. I ended up getting the President of Continental, Lester Miller, on the phone. Lester said, "I need to find someone to do the manufacturing."

That's when I said, "I want to do the manufacturing."

He asked me when we could start production. I informed him we were moving our shop to a much larger building in Vernon and would be in production in three months (a big lie).

The reality is we knew it was going to take more than three months to get our new factory set up and get into full production. Also, we needed the right presses to do the job and we didn't have them. So why did I make this promise you ask?

I knew that the job could make us.

But the problems were enormous. The first problem was we knew we had to have at least one big press, several smaller presses, and several pieces of related equipment. We studied the problem and concluded that the job was going to cost us well over $100,000 of investment money.

I went crazy.

The first thing was the big press. Buying a new press would have cost $90,000 – that was out of the question. Our best customer was Travis Industrial Electric. We were doing a lot of work for them, and they were employing about half of our guys, and they paid us good money. One day the Travis engineer was in our shop telling me about the troubles they were having with a big press. I was all ears. We ran down to their shop to find out what was wrong with the press. The engineer told us what had happened. The press had had a major breakdown. The company had hired a press repair company to fix it. They repaired the press but when they had finished with their work and fired it up, it still didn't work correctly. I became even more interested…

We ran down to their shop, climbed up the ladder to the scaffold that ran along the top of the press, so we could look down into the machine. After a few days of this, we figured out what was wrong – the crankshaft was 12 inches in diameter, and it drove another shaft which was 9 inches in diameter, and they weren't joined right. We went to the Travis brothers who knew all the problems and negotiated a purchase price for the machine for $15,000.

That was exactly the break we needed, but we STILL needed to fix the machine.

The press had to be removed from the building with two huge cranes because it could not be moved around in the factory itself. First, we had to build a foundation for the press which we were able to build in three weeks. After the foundation was in, we pulled the press out of the roof of their building, set it down on a huge truck,

The BIG 250 Ton Press that Made the Mop Wringers

moved the truck to 51st street and used one of the cranes to move the entire press onto the foundation. Uncle Dave hooked up the power to the machine. I rented a 20-ton forklift and we started disassembling the press. We knew exactly what we were looking for and we found it. It was a six-inch square key that had been put in backwards. This took two weeks to correct. We had to take the entire press apart and re-assemble it to fix it. Not only was the key turned around backwards, but it had also been made off set. We made a new key which took us an additional week. After putting it together, we fired it up and it worked perfectly. I almost passed out.

I would like to say that it ran perfectly from then on for 30 more years.

Now we had to solve the problem of having the entire factory ready. By the time we had acquired the above press, early 1965, we had already moved into 51st street, which was an ancient World War I, 30,000 square foot building in Vernon. We called it Building One. We had needed to remodel the building and do a lot of electrical work, so again I called Uncle Dave. He wired and re-wired the entire building which took almost six weeks. During that time, we bought four new punch presses and four used presses, got them all hooked up and into production. I also almost went broke! We had used up all our cash, but Union Bank came down to the shop, looked over our books and the manufacturing facility, and gave us a $200,000 credit line.

So even though I had promised Continental three months, the reality was with all the remodeling of our factory and the acquisition of the required presses, it actually took us 6 months. At that point we were making 400 mop presses a day, but Continental was expecting 500 a day out of us.

The problem was with the guys I had working on it.

We had moved in the 20 employees we had from the little building on Jesse Street and were using them for the repair of the big press. Two or three of those employees were being used for the stamping of the mop ringers. They were old reliable employees.

We had to hire several new people to assemble the mop wringers. My old reliable guys came to me and said, "The guys you hired are fuck-offs".

Our 22 Employees That Moved With Us to 51st Street

So, I spent a week out in the shop watching the assemblers. And one fine day, I called two of them to come into my office. "Wow," they thought, "we're in the big shot's office!" I explained my problem to these two fellows and told them that I had studied the workflow in the shop and had come to the conclusion that if the assemblers were accurately used, that they could without question produce the number per month that Continental was waiting for. I told them that they were going to be appointed Foremen of the Assembly and they had the ability to fire anybody that they thought was not putting out. I also told them that those two guys were going to get a small bonus for everything that they produced over schedule. They loved it! Within a few days they had fired and replaced several of the assemblers and they remained the Foremen of the Assembly of not only the mop presses but all other jobs that required the same techniques. This went on for many years.

My old reliable guys had come through for me and had gotten production going to 500 per day. And not a minute too soon! I had already made the gamble of my life, and nearly went broke on this deal. I HAD to make this work!

Fortunately, Lester and I were friends, and he knew all that we had put into this deal to make it work. Soon enough, Continental was getting the product they needed and was very happy with us, so much so they would at times pay us in advance. We then continued to increase production and sales to the tune of $120,000 per year for over nine years. And by the end of 1966 we had gone from 20 employees to over 60 employees.

The long and short of it was that the greatest gamble of my life had paid off and paid well!

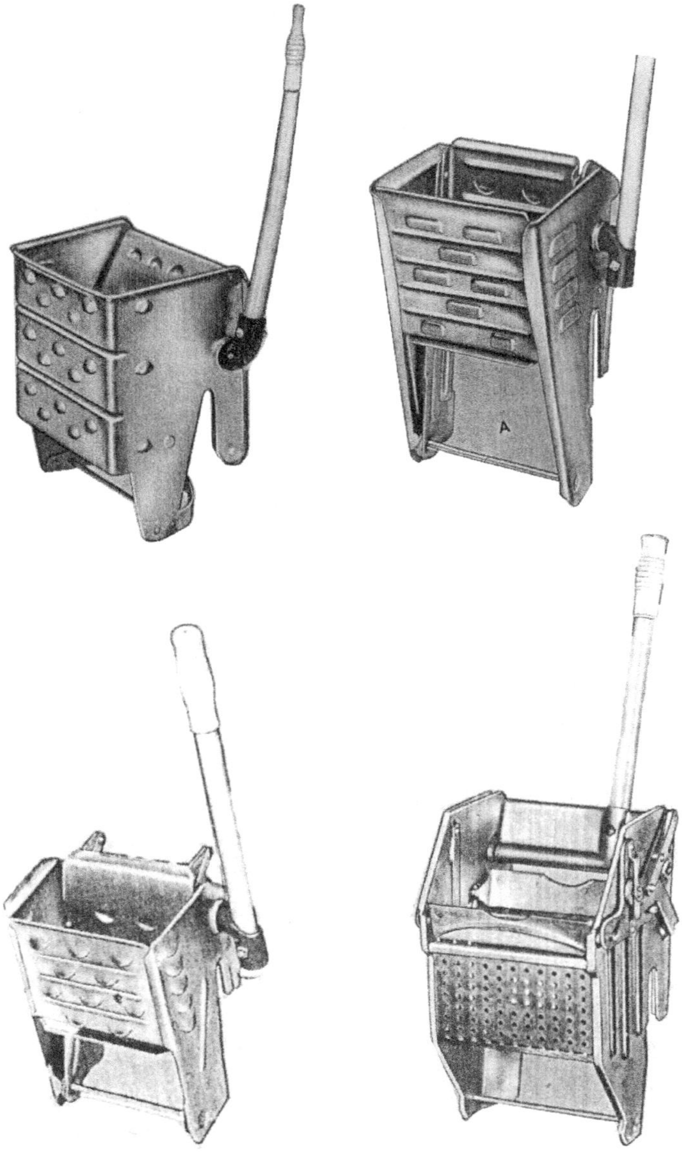

The Four Mop Wringers We Manufactured for Continental Janitor Supply
We made almost a million over 9 years!

CHAPTER SEVEN: Adelyn And the Expansion

About a year after the start of our mop press manufacturing business, in 1967 Shirley and I were at a Sunday afternoon Bar-b-que at the home of Bob and Shirley Knutson. Bob had become a close friend of ours several years before and knew everything about our shop.

Bob said, "How are you doing now that you've become a big manufacturer?"

And I replied, "We're doing really great in business, but I'm going nuts in my office." He asked what was wrong.

And I replied, "I simply am not set up to be the manager in a seven-person office of babbling women. Del and I both have a lot of work that we must concentrate on, and I for one am constantly interrupted by these babbling women. I need an office manager."

Bob turned to his Shirley and said, "Adelyn Miller."

Shirley said, "You're right. Give him her number." She told me that if I could get her, she would solve my problem.

I phoned Adelyn the next morning, talked to her briefly and she told me she would like to come to my office. I asked her when. And she said, "now." She did show up in a half hour. I greeted her at the door and took her in my office. We had a long talk. She was a grey haired, slender woman who was a retired physics and upper math teacher at Compton College. She also told me that her husband was a retired Vice President of IBM and they had originally come from San Francisco, where they owned an apartment house. They had to move to LA because of her husband's work for IBM. I explained my problems and she said that she would like to spend some time here and suggested that I announce to the staff that she was a consultant to find out about our procedures. She also told me that it could not be a full-time job for her because she had to have a lot of time off occasionally. She said that we could work that out. Adelyn came in the next day, and I introduced her to all the women in the office, and to my nephew George and our bookkeeper, Hiro.

The following week she knocked on my door, which was always open and asked if she could come in. I motioned that she come in and take a seat. She wrote down all her suggestions, which I agreed with. Two of the suggestions were the replacement of two of the women, with two other ladies whom she had met at prior jobs. We also talked about her wages. She suggested that she would be pleased to receive 60% of what a qualified office worker would get and told me that the two women that she suggested I hire could be trained to replace her during the week she wanted to have off once a month. Her two daughters and son lived in San Francisco with her three grandchildren, whom they wanted to visit once a month. They had kept a small apartment for themselves in San Francisco.

Within three months, Adelyn proved to be a goldmine for me. I loved her. She was almost old enough to be my mother, and at times I felt as if my mother was working for me.

For instance, one fine day, she knocked at my door, and I motioned, and she sat down and said, "Seymour, I have something to tell you that's none of my business."

I told her, that "anything to do with this office is your business."

She said, "Please excuse me, but I have been snooping around the office and I found the large storeroom across the hall next to the lady's bathroom. I noticed a lot of metal objects there, as well as many file cabinets filled with old records." She told me she called a plumber friend of mine, who came over. She showed him the room and the bathroom and told me what her suggestion was. She said, "If you can clean the storage room out, at a very reasonable price, we could plumb the room into a canteen." I asked her what a canteen was. She told me it is a little kitchen fully equipped so that we could have a sink, a tiny stove, and a refrigerator so we could have lunches there and an occasional cup of coffee. She said she had also called Hank, my carpenter, and he suggested that he could put up a table that was almost eight-foot square, where Del and I could spread our blueprints out and have a place where we could work in the afternoon. She said she noticed that we two fellows did not have a place where we could spread our stuff out.

I carefully paid attention to what she said, walked into Del's office, and told him the story. And he said, "Great. Do it in a hurry." This was a stroke of genius for Adelyn.

In 1968, our employment was up to 60 people in the shop and seven people, including Adelyn in the office. We were busting at the seams and had filled both buildings. The real estate salesman from DAUM Industrial Real Estate that we had done all our business with starting with the lease and then the purchase of Jesse Street, and who had sold us Building Number One, had stayed in touch with us.

One day he came to see me and said, "You obviously need more space."

And I said, "You're right, we need it desperately."

He said, "I have the answer to your problem, which is directly across the street. Let's go across and you can look at the property." He opened the door to the building across the street. We walked in and I had never seen such a mess in my life. He said, "$150,000 buys the whole mess. That includes 20,000 feet of a worthless four-story ancient flour mill standing on huge I-beams. Next is the 30,000 square foot warehouse, and outside is over an acre of parking space next to Alameda Street." He told me they would take $50,000 down and the balance payable over two years.

Del and I looked at each other and simultaneously said, "Sold." Our company bought that property, Building Number Two.

CHAPTER EIGHT: The Split

One day in 1969, Del walked in my office and sat down. He said, "I have great news." I asked him what that was? He told me "I'm going to marry Julie." He brought her to the office to introduce her to me. It happened to be a Thursday when Shirley was working. Julie was a young blond blue-eyed Hollywood chickee type that Del had been seeing for months. Shirley came into my office, interrupting her payroll job. As usual, Shirley was incredibly quiet with strangers, and this stranger was almost young enough to be her daughter. I did not know it, but this was going to be the first time that Del and I had a problem with one another.

Del bought a little house near Orange County where Julie had grown up. Julie also wanted a fancy little car, which Del bought her. All went well, till one day, Del came in and said, "I want Julie to come in here and work in the office the same as Shirley does." I told him that I had no idea what she could do here. So, I told him to talk to Adelyn, which Del did. Julie started coming in a couple times a week. One day, Adelyn knocked on my door. I motioned for her to sit down.

She said, "Seymour, again I'm in an embarrassing position." I asked her what the problem was. She got up and closed the door and sat down and she said, "Julie". I asked her what the problem was with Julie. She told me, "I can only tell you that as the office manager, which you say I am, I'm telling you that she is absolutely worthless here. I cannot say anything else other than that she is a troublemaker." She stood up, said she was sorry and turned around and walked out.

The next day I asked Del to come into my office. Before I had a chance to talk, Del told me, "A couple of months ago, before we were married, I told Julie that the company had donated $50,000 to the new Music Center in downtown LA." And then he proceeded to tell me that Julie had told L. Ron Hubbard, the head of the Church of

Scientology, what we had done. And Hubbard had told Julie if her husband had done a silly thing like that, then he would expect her husband to donate at least $100,000 to the Church of Scientology. Also, he continued to tell me that in her visits to the office, Julie had found out that Shirley signed the payroll checks and other checks.

I sat quietly and finally I asked Del, "What is this leading up to?"

Del said, "I own half this company, and I want my wife to have the same ability as Shirley." I asked him to tell me exactly what he wanted. He said, "I want to donate $100,000 to the Church of Scientology, and I want Julie to sign on our bank accounts."

I sat there dumbstruck! And finally, I told him that would never happen. Del got up and walked out.

A few days later, on a Sunday evening, I received a phone call from Del. He said, "I'm at LAX and I'm waiting for a flight to London."

I said, "London, what's in London?"

He told me Julie was in London, living on Hubbard's barge on the Thames River and said, "I'm flying there to bring her back."

I said, "Holy cow! How long are you going to be gone?"

He said he did not know but he would let me know. I did not hear from him for two weeks, during which I had to take over Del's duties in running the shop. I was working twelve hours a day, seven days a week. Finally, I received a call from Del. I asked him where the hell he was. He told me he was down near Palm Springs sitting under a rock. I asked him how long he was planning to sit under a rock. He told me he would be back within the month.

In 1968 our agent at DAUM called me up and said, "Seymour, I have more good news for you. The printing business next door to Building No. 1 has gone bankrupt, and the owner of the property is going crazy because of lack of rent and wants to get out. It's 40,000 square feet of building adjoining Building No. 1." I asked him how much. And he said, "$200,000, with half down and the balance over two years." I told him we would take it. My agent from DAUM asked me if we wanted to see it and I told him whatever it is, we will want it. I asked him to come on down to show it to us. Sure enough after we saw it, we bought the building right away.

In 1969 we purchased another old ramshackle property at 2300 E. 11th Street in LA. We had almost completely run out of room again and we decided that we were going to move CTD Machines to the new property and put my nephew, George, in charge of that company. We had a lot of work to do to clean the building up, which we started immediately.

In that same year, Adelyn came to me and suggested that we hire a girl that she had known for 10 years. Her name was Nancy Sheek and she was 22. She had just graduated a college accounting course with high grades. We needed an assistant for our accountant, and she knew that Nancy could do the job. I hired her and she did do the job extremely well.

Soon after we had hired Nancy, Del suddenly showed up in my office after having been gone for about six weeks. I asked him what the hell had happened? He said, "What happened was that Julie had taught me things about myself that I didn't know."

"Like what?" I said.

He said, "Let us face it. You own this business."

I sat there quietly and then I asked him, "How can you say that I own this business?"

And he said, "You run this business as if you owned it."

And I told him, "You own half the business."

And he said, "Julie taught me that if I owned half the business then I have the right to do what I want with my half."

I quietly told him that if he wanted to take his money, he could do anything he wanted with it, but his wife does not have the ability, nor does she have the right that my wife has. I told him that Shirley had been a signer on our account since I opened it--and so had Pat (Del's ex-wife) and he knew that. Del said, "I did know all that, but I want to be an owner of my business and not have to answer to you." I quietly asked him what he was proposing. And Del quietly said, "I am demanding that we split the business up and part company." He then got up and walked out. Shortly thereafter, he informed me that his marriage was being annulled.

For the next couple of weeks, I wracked my brain and had several other little sessions with Del. Here is the result: This is a copy of the agreement that Shirley typed out. We did not want anybody at the shop to know anything that was going on.

AN AGREEMENT TO PURCHASE PUNCH PRESS PRODUCTS AND PART OF COMMERCIAL TOOL AND DIE, WHICH APPLIES TO EACH PARTNER

- CTD Machines, Inc. and the real estate do not apply to this agreement.
- The purchase of one-half interest of Punch Press Products and Commercial Tool and Die--$1,000,000.
- The down payment is $300,000 in the year 1970.
- The balance of $700,000 is payable over ten years at 8% interest, with minimum payments of 10%.

- In addition, the seller will receive rent of $22,000 per month and 10% of the profits of the company for ten years.

The next day I gave Del a copy of the agreement. He told me, "I was under the impression that we were going to split the company."

I told him that I could not figure out how that could work and keep the company going the way it is. Del told me that he accepted the agreement and that he wanted to buy me out. But where the hell was he going to get $300,000 cash? And he wanted to know where I came up with the $1,000,000 price.

I told him, "Del, do you remember after we got home from St. Louis, and we told Lester Miller we were not interested in the merging of our company?" (Lester had at one point offered us an opportunity to merge with Continental Janitor's Supply) He told me that before they invited us to go to St. Louis that I thought we were going to do close to $2,000,000 in 1970, and for sure make that much in 1971. Lester told me that if we did $2,000,000 of business and made at least 10% profit, then the company was worth $2,000,000. And that was the reason why a month later, after we had turned him down on the merger, he called me up and offered to buy 20% of the company for $400,000 cash on the barrel head.

Del said, "Yeah, I remember that, and it makes sense." We came to an agreement on the $300,000. I then received a $100,000 deposit, and the balance of $200,000 at the end of 1970.

I prepared a list of the machinery that I wanted to take from Commercial Tool & Die to move to 11th Street. Shirley and I had become friendly with Nancy. I called her into my office, and I told her, "I'm going to tell you something and I want you to keep it strictly confidential. Shirley and I are leaving Punch Press and we are moving to the 11th Street property, which I am going to own. And I am offering for you to go with us. You must realize that you will probably have to spend a lot of time in the office with nobody around. But after a while we will probably hire you an assistant. In the meantime, Shirley will be working almost full time in the office." Nancy said that she accepted my proposition and mentioned that it would be nice having a quiet office.

At the end of 1969 right before the final split, I moved CTD Machines to 11th Street. The office was completely equipped with old furniture that I talked the prior owner into leaving. For the year of 1969, CTD Machines did $125,000 of business and made a slight profit. By the end of 1970, CTD was up to $200,000 in sales and was making a nice profit. Prior to this time, Commercial Tool & Die did all the machining, and the machine company only assembled the machines. On 11th Street, we did all the machining as well as the assembly.

Del lived up to the letter of the agreement, made all the payments and in 1972 did over $2,000,000 worth of business. During the next ten years, his company rapidly gained in sales and profits, and at the end of ten years his sales were at the $10,000,000 level, and he had 125 employees. Del was happy to have paid me off and to be able to stop paying me 10% of his profits. But he did not like the idea that he had to pay me rent, which had grown to $35,000 per month. He told me he wanted to buy my half of the buildings. I sold the buildings to him for $1,100,000, which was the current value at that time.

Fortunately, Del and I remained close friends throughout the remainder of our respective careers, and I still feel so fortunate to have had him as my partner all those years.

Del and I Remained Close Friends
This Picture Taken in 2012

CHAPTER NINE: The Development Of The CTD Inc Line Of Machines

There was another whole aspect of the growth of Commercial Tool & Die, which in order to illustrate, I need to take the reader back to 1956.

In the middle of that first seven-year poverty-stricken period of Commercial Tool & Die, Milt Gray showed up in my office. Milt and my brother Dan both had worked in the engineering office of Douglas Aircraft for years. After the war, Milt was able to take over control of Thackaberry Tool & Electric Supply Company. The company had been formed in the World War I, and Milt's family was a partner in it.

Milt said, "I'd like to go through your shop and see what you have." Later, we sat down in my office, and he said, "Here's my problem. I am doing a great job of selling light machinery that needs to be adapted to other machines. We do not have any capacity to do any machine work and I see that you can build just about anything here. What I'd like to do is give you work on a time and material basis and see if we can get along." That started a twenty-year relationship. We did several jobs for Milt.

In 1959, Milt called me up and told me to come to his new location, which he had built in a beautiful new building not far from our shop. I went there and he introduced me to Edgar Pope. There were two machines laying on the counter. Pope had just gone broke, and Milt explained to me that Pope had to sell out as quickly as possible. The machines were explained to me in careful detail. I called Del, my partner and asked him to come up to Milt's store. We explained everything to Del. We took the literature on the machines, and we went back to our shop.

The next day we told Milt we would pay $2,500 for the business--$750 down, and the balance over three years. Milt got back to me later and told me the figure had to be $3,000 and I agreed. We brought everything back to our shop, which included all the aluminum casting match plates. That started Commercial Tool & Die Machines, Inc, which later was called CTD Machines, Inc.

We developed our first line of machines which we called the P-100's. They were used to cut primarily wood but also aluminum for cabinet work that required miter cutting. We only made the base; the motor was an electric saw made by Portal Cable.

P-100 Our First Line of Machines

Incidentally, it was during this time when I was developing the P-100, that Carl Weitzel had come to survey my shop and had recommended that I go into the manufacturing business, which of course I eventually did.

Also, as I was perfecting the P-100, we had bought another machine with the identical base, but it had its own spindle assembly and conventional electric motor. I worked on the original machine, redesigned it completely, improved it greatly, and we called it the S200. We then developed the M225 from that machine.

I believe we sold well over a 1,000 of these machines.

There were other models of the 200 Series machines that we constructed. One of them was called the DM200, which was a right- and left-hand adjustable machine that rode on a ten-foot-long base and could cut both ends of an eight-foot-long piece of wood or aluminum. The machine could miter and was completely adjustable, and just the right machine for window manufacturers.

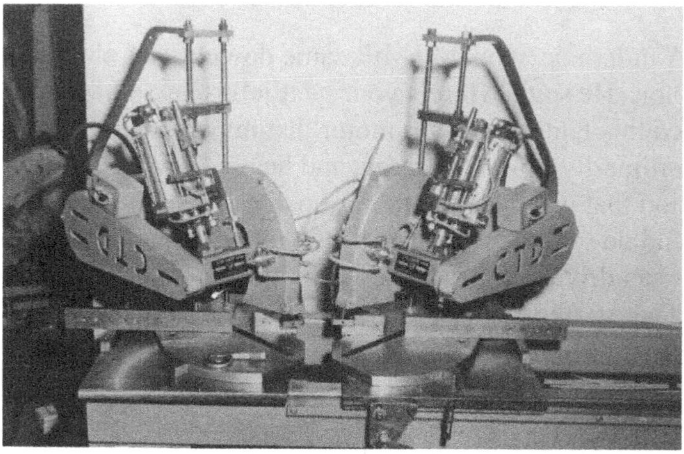

DM200 Double Miter Chop Saw

CDM200 Compound Double Miter Chop Saw

A month or two later, Milt came down to our shop lugging a huge motor. He said, "Here's your next job. The Stanley Works has developed this fantastic router motor that nobody can use because it is so extraordinarily powerful, large, and heavy. I want you to make a contraption that will pick the motor up 6" with an air cylinder and will ride up and down on a couple of heavy rods. It must be strong enough so that it can drive a router bit straight through a pile of wood in an instant. That started our few years in the routing machinery business, and with sculpt-a-grill.

What is a 'sculpt-a-grill' you ask? Let me explain: After the war there were thousands of small houses being built that people wanted to doll up. They were taking particle board and cutting holes in them with screwy shapes and designs and folks were buying it like mad to decorate their houses. The problem was that they were doing it with small hand routers, which took a long time to do the job. Larger customers wanted a machine to do the job. In the machine tool industry, there was a technique called tracer valve machining, in which a wood worker set a template on the machine and adjusted the tracer valve to machine the stack of plywood according to the template.

Milt said, "I will furnish you with the valve and the pump, and all the electrics and router motors, and you build the machine. Except you have to give me an idea of what the machine is going to cost." He gave me an order. We built the machine. It worked and I then built a machine with two router motors that worked at the same time.

A couple of months later, Milt came to me and said, "US Plywood in Lebanon, Oregon has asked me to quote on a machine that could route out the thick plywood bases of General Motors trucks and buses. They would have to cut a hole big enough for the transmission and the wheels that extend up through the base. And they want the machine to be able to have four feet of cross travel and sixteen feet longitudinal travel."

I said, "Holy Cow, that's going to be a monster of a machine."

Milt said, "You gotta bid it right. My part of it is going to come to over $15,000." I thought about it for a week after I designed the machine and told Milt that my part was going to be over $20,000. It was going to be a four headed machine. We got the order. It was so big, we had to build it outside, in our yard under the Sixth Street bridge. It took us four months.

1962 GM Truck US Plywood Router Machine

US Plywood sent two representatives down from Oregon. We ran the machines for them, and they bought it. They told me that they were going to send down a big truck and trailer and that we were going to have to take this thing apart, load it on the trailer, and send somebody up to Lebanon to put it together.

CTD's foreman was Joe Perez, who had worked for me for several years. He was a crackerjack machinist and welder, and a hard worker. Joe knew that part of the order included him having to go to Lebanon, Oregon to put the machine together. I told Joe, "I will fly you up there."

The next day Joe came into my office and said, "My wife told me that I wasn't going to fly and that she wanted to drive up there with me and see the country." And that's what happened. Joe started the trip, and it took a week for him to get to Oregon, and by that time the machine was waiting for him on the truck. He and a couple of US Plywood workers put the machine together, which took almost two weeks. The last time I heard about the machine was about ten years later, when one of the valves failed. We got it fixed. I believe the machine is still working sixty years later.

I also decided to make what we called the 400 Series that had a larger spindle that could accept a larger blade than the 10" or 12" 200 Series. The 400 Series could accept motors up to 20 horsepower—with up to a 20" blade.

Trade Show Featuring Several of Our 400 Series

After a couple of years, I designed what we called a 600 Series spindle for blades up to 36" and designed for cutting aluminum billets and abrasive cutting of larger sections of steel. We sold a 600 Series to a company that wanted to cut up 8" diameter aluminum billets. It was highly successful. Next, we used the 600 Series spindle to make several huge abrasive cut-off machines. These used 50 and 60 horsepower motors.

One day, around 1967, Union Pacific Railroad came to see us, and I showed them the spindle assembly on a large machine I was manufacturing for another company, and they said that they would be interested in seeing it work. A few months later I called them to see the machine I had built, and they came down with some pieces of railroad track. The machine had a 50-horsepower motor and turned a 40" diameter abrasive blade, with a reciprocating motion. The blade moved front and back while it was sawing through the material. They bought it on the spot. It cost $6,000. We started the construction of two more identical machines, finished one of them and Union Pacific bought it.

One interesting note is that anytime I designed a new machine, I would come into the shop and complete all the tooling for it. This would surprise many of our shop workers because most of the time, I was busy supervising the production or working in the office running the business side of things. They would often say to me, "I didn't know you were a machinist; I didn't know you were a tool maker." Yes, I would do all the initial tooling, then I would hand over the prototypes to my foreman to fulfill the production.

In 1968 we had manufactured quite a few machines and were in the process of crowding ourselves out of the shop. In 1969 we bought an old property on 11th Street in LA because we were being forced out of 51st Street by Punch Press Products, a sister corporation, due to the fact of Del and I splitting the companies.

In 1970 after my split with Del, I designed the D45 which was a unique double miter saw where two blades worked simultaneously to cut picture frames or any other material that had to look like a picture frame. It was a completely new design that I had been thinking about for two years. It took me more than a year to develop it, but in the end, it turned out to be our biggest winner. We shipped them all over the world.

D45 Double Miter Cut Off Saw

The company then continued to enjoy more stability, continued growth, and profits for many years.

CHAPTER TEN: Ellen and Tom

As I mentioned in a prior chapter, when Del and I had split the companies, Nancy, who had been discovered by Adelyn, moved with us to our new location and helped us successfully run the office for seven years. Adelyn had decided to stay with Del.

In 1978 Nancy told me that she needed to quit because she was pregnant. At about the same time, I was able to hire my daughter Ellen to replace her. Nancy trained Ellen, who was a fast learner, in our procedures, except Nancy neglected to tell Ellen about credit problems.

One fine day, an order came upstairs with a note that the machine, a DM200 was ready to be shipped but it came along with a note questioning as to who would be coming to pick it up.

Ellen said, "We're shipping to Oregon by such and such company." She did not realize that the machine had not been paid for, and out went the machine. When I came back to the office, Ellen put the copy of the invoice that she had made for $2,200 on my desk.

I looked at it and said, "Wait a minute. That machine was supposed to be paid for in advance." The problem was that there was a little shelving unit on the wall above Ellen's desk and any order that had problems with it was kept on the bottom shelf. Ellen did not know that.

I told her, "You should not have shipped the machine." And she told me that nobody had told her that. The invoice was sent and a week later I called the guy in Portland, and I told him I wanted his check. And he said he would send me a down payment and pay the balance in a couple of months.

In the meantime, I had chewed Ellen's ass out and she sat on my couch and broke down in tears. Do you know what it is like to have your daughter sitting on your couch in the office crying her eyes out? I called my lawyer who referred me to a lawyer in Portland. The bottom line of this story is that he collected all the money and charged me $400 to do the job. What happened was that the $400 was the best money I ever spent because it turned Ellen into a credit gorilla. From that day on, she did not trust anybody, and she turned into the greatest accounts receivables collector that the world has ever seen.

A year later, Tom Orlando, Ellen's husband went to work for me as the sales manager. More than once I would be sitting in my corner office listening to Tom and Ellen arguing about business matters--Ellen was saying things like she "knew that son of a bitch and she would not trust him for a dollar. And if you want to sell it to him, he is going to have to pay for it in advance." It was a $4,000 machine that poor Tom had spent two weeks trying to sell. I am sitting in my office, listening to all this, and I am thinking that I must have died and gone to heaven.

Tom and Ellen Orlando

Ever since the split, I had determined that I wanted to invest in industrial properties, and only industrial properties. I would buy shit properties, fix them up and rent them, and did this about twenty times. I would sell them at a profit and use the money to buy better properties.

The bottom line of this is that by the year 2000, I had bought 49 properties, sold 36 and had 20 tenants in 14 properties. And I still have the same to this day. The only difference is that Tom has become my Trustee and the official in charge of operating and managing my properties.

In LA, Ellen was my bookkeeper and took care of my rent receivables and paying the bills. We did this for many years. Ellen and Tom worked for me for twelve years and bought CTD Machines in 1990. At the same time Shirley and I moved to Montecito and I took over all the bookkeeping for all the rental properties. Ellen and Tom continued to stay in LA and run CTD Machines for several more years.

CHAPTER ELEVEN: My Lifelong History with Daum Commercial Real Estate

In 1904, William H. Daum and his son, Bill Daum Jr. formed the DAUM Commercial Real Estate Company. His other son, Richard Daum, became part of the company about 20 years later. Mr. Daum's emphasis was confined only to manufacturers and distributors—primarily in Vernon and new surrounding neighborhoods that were developing. But one of the most important things that they did in my opinion, was to solicit purchase agreements from disgruntled farmers that had land that was not very productive. They received option agreements at a fixed price and date. It is my understanding that in the forty-year period up to the end of World War II, the DAUM company brought over 3,800 companies from the Eastern parts of the United States, Canada, and Mexico to the Los Angeles region. Some of these companies are the largest in the world. Mr. Daum passed at the end of the War and his two sons both passed in the 1970s.

In 1954 I called DAUM from a sign. They sent down a young salesman named Bill Davidson. I explained to him that we needed to lease a small cheap building to be used for a tool and die shop. Three days later, he came back to see me and my partner, Del Molinari, and we followed him down to Jesse Street. It was half of a 5,200 square foot building. We walked in and immediately saw that all the walls were black. Bill told us that we were lucky the building came like that. The walls had never been finished on the inside. As a result, we'd be able to rent the building for almost half the amount that it was actually worth. He said the rent would be $150 a month, which would have been $.06 a square foot, so we'd be saving $100 a month, and if we put drywall up, we'd have white walls. We loved having the savings of the $100 a month. And eventually, we did put drywall up. We had a three-year lease, with an option of three additional years at the same rent. Six years later, Bill represented the seller and sold us the building. That was 1959. Within three years, we bought three other small properties—one of which we used, and the other two we leased out.

In 1964, we went into a terrific expansion and bought a 28,000 square foot old building in Vernon. Two years later, we bought a 30,000 square foot ancient warehouse across the street from the Vernon building, which had a large parking lot. Three years after that, we bought a 40,000-square foot adjoining property to our first Vernon property. In the early '70's, Del and my terrific partnership split up. I wound up with the smallest of the three corporations, which was a manufacturer of small industrial machinery and moved to the building on 11th street which Del and I had originally bought. In the split up, I wound up owning all the property with options to Del to purchase the properties that he occupied over a ten-year period. Del became my tenant. As I described in the earlier chapter, I received a handsome payout from Del.

By the end of the '70's, I had acquired approximately 20 properties, almost all of which were handled by DAUM. I was called to the DAUM office for a meeting. A group of DAUM people were in attendance, and they wanted to explain a plan that they were working on. The plan was called Checkmate. It consisted of the rebuilding of certain ancient areas of Vernon. Many busy streets were lined with old buildings that had no parking and no resources for loading and unloading, and their owners were begging for tenants. They were slum properties. The plan was to purchase the properties that were adaptable, tear down adjoining properties for parking and loading, and sell them as improved properties. The DAUM men told me that they had a plan that would be strictly a DAUM proposition, and they needed outside money for financing the plan. I was not interested in financing the plan, but eventually bought two properties, improved one of them according to their plans, and sold it two years later. The second property, which I still own, got an easement for loading and unloading.

In 1983, DAUM was sold to an exceptionally large Washington DC based construction company with the understanding that DAUM was going to be the West Coast headquarters for a national industrial real estate brokerage firm with offices throughout the country. The buyer was a five-billion-dollar company and gave the sellers the instruction that they would continue to be the managers of the company and would provide DAUM with any amount of funds to purchase other companies in the area. The number of salesmen at DAUM rapidly increased to over 300 in three years.

In 1986, the parent company went bankrupt, and DAUM became a small part of a huge bankruptcy estate. The bankruptcy manager sold DAUM to a major Dallas based construction company, which in 1988 also went bankrupt. Then the company was sold to two land bankers in Georgia who were not interested in operating DAUM. By that time, with all the bankruptcies, more than half of their people had quit, knowing that DAUM itself was going to go down the tubes.

In 1989 Mike Nubel, who was my exclusive contact at DAUM, called me to arrange a meeting in my office in LA. Eight people showed up, many of whom I knew. They announced that they wanted to buy DAUM, and they wanted me to finance the deal. They were each going to put up $100,000 which would buy them 10,000 shares at $10 a share. I told them that I was willing to buy $100,000 worth of shares, but I was unsure about putting up the money to finance the company. We did end up purchasing the company with a large down payment and the loan being financed by the sellers. The nine of us then hired someone to be the manager, and Mike stayed in constant contact with me about the company and its operations.

In the middle of 1990, Shirley and I moved to Montecito. My conversations with Mike continued and he was complaining about the new manager who had only been there for six months and did not know what he was doing. Additionally, the company desperately needed more money. None of my partners were willing to come up with extra money. They were in a jam, and I told Mike I would agree to put up another $100,000 on the condition that I would take over being the manager. Mike came back to me and said that the partners decided to actually make me the Chairman of the Board. I then told Mike that as the Chairman, I was going to fire the existing manager and make Mike the manager. Mike and I ran the company for three years at a loss. I steadily put more money into stock.

Finally, around 1996 the company turned a profit and the same in 1997 and 1998. At a Board meeting, the partners started squawking that they had made an investment, making no income for almost ten years, and that seeing as how the company had made money for three years straight, they wanted dividends. By this time, I was by far the largest investor, and I would have received a considerable income from the dividends. But I insisted that the company could not afford it, and that we were building for the future. Eventually I had to give in and allowed dividends to be paid. Mike advised the Board at our monthly meeting that there were several very profitable salesmen that wanted to own stock in the company. My partners did not want to issue more shares because it would have diluted the value of their shares. By that time, my family and I owned 93,000 shares, out of 150,000 issued shares. It was finally agreed that the company would issue 50,000 more shares, and I realized that this would cause me to now own less than half and lose control. My partners, however, agreed to sign a contract stating that they would never oppose me having control of all the shares. Over the next fifteen years to the present, the company has prospered and has delivered extensive dividends to the original shareholders, as well as the new owners. And that is where we are today.

Starting with the purchase of Jesse Street, which was engineered by Bill Davidson, several DAUM salesmen had sold me properties. Larry O'Brien was the second. Mike Nubel was the third. Several others also participated in my purchase of 49 properties, plus I entered several partnerships.

Over the next forty years, I spent a lot of my time repairing and selling most of the properties, and at the present time I own 15 properties with 20 tenants. While this was going on, I described all these deals to Ellen and Tom, and they started to realize that maybe this is what they wanted to do as well.

The reason for this extended illustration is to explain my entry into the industrial property business, including what had happened with the formation of DAUM over the years.

PART II

CHAPTER TWELVE: My Love Of Nature & Cabin Life

My lifelong passion for being in nature started with my parents who loved the mountains and hiking. One fine day a couple of years before I was born, they discovered Big Tujunga Canyon (Big T) in the range of mountains North of LA. They were driving on the dirt road and came to a little sign pointing up towards a canyon, that said "Trail Canyon, Cabins 13 – 30". They continued driving up the little dirt road and came to a parking spot. They parked and started walking up a trail and entered the canyon. The right-hand wall was perhaps 200 feet high and extended as far as you could see. At the base of the wall was a little stream winding to and fro through a group of trees that were all over the canyon. The stream went under the trees and in places it was flat, and in other places there were little pools. Next to the stream was the dirt road.

My parents loved the sight. They walked up a small trail and found a little crappy cabin with a 'For Sale' sign on it. The sign pointed to the next cabin and said, "For information, go to Cabin 14." They looked all around the cabin and went to Cabin 14 and knocked on the door. An elderly lady answered the door and they told her they were interested in the cabin next door. The lady told them that the neighbors were good friends of hers and had been trying to sell the cabin for several months and they had reduced the price to $50. The lady asked them if they wanted to see the inside and they nodded yes. She took them over and unlocked the door. There were four beds, an old iron potbellied stove, a little table, and a few chairs. That was all the fixtures of the cabin. A couple of weeks later they came back and were snooping around the cabin. They found a water faucet and an outhouse about 20 feet away. They went over to see the lady and told her they wanted to buy the cabin. The lady told them that she would make the arrangements for the sale.

The arrangements were made, and Ma and Pa brought their two sons, aged 13 and 10, Dan and Sam, and they all started cleaning the place up. They also started bringing all the little things that were needed, such as bedding and kitchen supplies. This went on for three or four years, during which time I came along in 1927. Ma and Pa loved Trail Canyon. They hiked all over the place. The trail actually went twenty miles to Mount Lukens. Across the creek from their cabin, lived Old Mr. Baker (old was part of his name). Old Mr. Baker had a contract with the forestry department to supply two lookout stations, one on the way, and one on Mount Lukens with food and supplies. He had a herd of mules and three or four horses and had to go up to Mount Lukens once every ten days.

Starting when I was four or five my folks would take me up to Big T and I started my adventures in the creek. Ma would watch from a window, and I knew she was watching me, so I was being very careful to not fall in the stream, which I did constantly. I would clean out little paths amongst the leaves for the stream to run through and block other places with little rocks. I loved that stream. Ma would walk down, and I would show her the work that I was doing, and she constantly told me to be careful because I could fall in and hurt myself. I heard that a thousand times.

In 1932 my parents separated, but they still were very friendly, and I guess Pa gave Ma the cabin. Ma never drove, but my older brother Dan had a Model T Ford and used to drive the family up to the cabin. I would jump out of the car and run down to the creek. This went on for several years. Dan and his wife Anne would come up to the cabin with Ma and I and stay for the weekend. In the meantime, Ma decided to improve the cabin. She hired Jack McBow, who lived locally to electrify the cabin. He then added a bedroom and then a kitchen. Ma had friends that drove, and several times she would invite them to come up and stay the weekend.

After I met Shirley in 1947, I told her about the cabin. You have to understand that Shirley was 18 and I was 20 and I had the hots for her! Hormones were in full flush, and in the back of my mind, I was thinking that the cabin could be the perfect place to fulfill my passions for her.

One fine day, I took Shirley up to see the cabin in hopes of making love to her. She was not impressed.

She asked, "Where's the bathroom?"

And I said, "We don't exactly have a bathroom. If you want to take a bath, we heat some water up and you take a rag and sit outside and wash yourself."

She said "Yeah, that's pretty rustic, but where do you go to a toilet?" I pointed to the out-house, which was now about 50 feet away. Shirley was even more UN-impressed!

From then on, every time I mentioned that maybe we would one day stay at the cabin, she would say "not me!"

Occasionally, after we were married, we would drive up to Trail Canyon and walk up the trail to what was called 'The Falls.' It was about a one-hour hike, and one could go above the falls, and have lunch in a beautiful spot, which I loved! Though Shirley enjoyed the hike with me, she never did agree to staying at the cabin.

The Original Cabin in Big T

Shirley in Yosemite on Our Honeymoon

Family in Sequoias, we spent a lot of time in nature as a family

In about 1956, Ma received a message that there had been a huge flood, and half of Trail Canyon was washed out, including the road into the canyon, along with our cabin. My brother Dan had called me to tell me the news. We drove up a few months later and I was able to walk down to see what had happened. It was really hard for me to take, and I almost broke down. From then on, every few years, Shirley and I and the kids would go to Trail Canyon and go on the hike to the falls, and I would walk down to the little spot where I used to play in the stream. Ever since then, there has not been a month that has gone by that I did not think about that little stream and all the joy I had as a kid playing there.

In 1970, my business was running steadily, and I realized I did not have to work so many long hours anymore. I began to reminisce about my childhood at the Trail Canyon cabin and I began to think about getting a new cabin that I could enjoy with Shirley and the kids. At the end of that year, I told Shirley "I'd like to drive up to Big T and buy or rent a cabin." We drove up to the entrance area and found a small real estate office.

I met Lou Miller who said, "Can I help you?"

I told him "I'd like to talk about getting a cabin in Big T."

He said, "well you came to the exact right place. I live in Jasper Flats and there are two cabins available."

We drove about 10 miles up to this little cabin. There was not much there, but we weren't exactly expecting much. There were two small bedrooms with a bathroom in between, a kitchen, and a living room. The rent was $75 a month for a year. We took it. We were there for a year before we decided to give it up and buy a larger cabin on the other side of the river. This cabin we paid $20,000 for. It had two bedrooms, one above the other. Off the upstairs bedroom was a large deck surrounded by a wood railing. We were there for four years. And then we were flooded out. Fortunately, we had flood insurance.

I told Lou Miller, “You have to find me a place above the flood line.”

And he said, “I have just the place for you! It’s right across the road from my house. It’s 40 feet or more above the highest possible flood plain. But the problem is, it’s on fee simple land, meaning that you own the land. You can buy the place for $115,000 and there is no dickering.”

We bought the house. We spent a lot of money fixing it up. We were there for several years, and I loved it! Shirley was able to select the new kitchen which we put in. The big bedroom had a private toilet and sink, and there were two small guest rooms between the bathroom. There was a huge rock fireplace at the center of the living room. Off the living room was a large, screened porch overlooking the creek, which was about 60 feet down the hill. It became my own little slice of heaven for about six years.

CHAPTER THIRTEEN: Our First Big Trip and Other Travels

During the summer vacation of 1963, we loaded our new three-year-old Cadillac up with enough luggage to last for several weeks. We then ventured north with our destination being the Port of Vancouver. We took our time and traveled for about three and half weeks, visiting both the Avenue of the Giants and San Francisco. In Washington state we then took the car ferry to Vancouver Island and spent a couple days there. On one of those days, we visited the famous Butchart Gardens. It was gorgeous! Absolutely unforgettable!

From there we took the car ferry across to the city of Vancouver. I drove up to the dock where the famous SS Oriana was docked. This enormous vessel was the Queen Ship of the P&O lines of England. Because we had driven our car, we had to store it on the ship. The attendant instructed me to drive the car onto a giant net. We then got out and went to the side. I watched while my Cadillac was being gathered up and lifted in the air by this giant net where it disappeared into the hold where they were keeping the vehicles. For a minute I was a little nervous watching my car being raised up in the air like that and hoisted away where I knew I would not see it again for eight more days. But I knew we were in good hands.

Five times a year, the Oriana sailed around the world, particularly to areas where there were British settlements. It would sail from London to Gibraltar, from Gibraltar to Malta, from Malta to the Greek Islands, and through the Suez Canal to Cape Town, South Africa. From there it traveled across the Indian Ocean to India, where it made two or three stops. Then it sailed on to Singapore, then Sidney, Australia, then to Auckland, New Zealand, and then to Vancouver. From Vancouver it sailed to San Francisco, then to San Diego, and then through the Panama Canal. It would then stop in the British Indies before heading back home to England. Passengers would board and exit the ship at all stops.

When we boarded in Vancouver, we checked into first class, as second class had already been filled up. At that time, to sail from Vancouver to San Diego cost $500 for two first class cabins. Also, interesting to note, in Vancouver, a large number of passengers would get off the ship and board onto the Canadian Pacific Railroad to go across Canada. In those days $500 was a lot of money but I decided we would do it. And I did it because I could not get second class tickets.

In first class, every five passengers enjoyed a dedicated waiter. We always ate at a table for four and shared our waiter with an adjoining table occupied by one very dignified lady named Mrs. Coldicott. Our waiter was Kith, we of course at first called him 'Keith', and he corrected us. His name was pronounced "K-ai-th" spelled Kith. Mrs. Coldicott was a British lady whose husband was a high officer in the British Army. She lived in Sidney and every two years, she would sail back to England to visit her family. She must have taken dozens of trips over the years. Our kids spent a lot of time with her and enjoyed watching the way she would cut a banana with a knife and fork, never once touching the fruit with her bare hands. She was extremely proper. We often asked her to dine with us but we would have had to sit at a table for six, and she simply could not imagine doing that. She used to take Ellen around by the shoulder after dinner, as Ellen was the youngest, only 10 years old. Karen was 13 at that time. The kids were fascinated by her.

It took us about eight days to sail from Vancouver to San Diego where we were finally dumped off. We drove back to LA from there. The minute we got home, the kids jumped out of our car and disappeared, only to return three hours later which they had spent telling their friends all that they had seen and done in the prior weeks.

SS Oriana

After the trip on the Oriana, I stayed very close to home for a couple of years. However, periodically we would take drives all over the western states and eventually over a period of about three years we visited every state in the Union except for North Dakota and Iowa. This involved several week-long drives, sometimes flying East, renting a car and driving for a couple thousand miles, then flying home. I was able to call on customers of CTD Machines that we had accumulated and had needed calling on. So, I ended up doing a fair amount of 'mixing business with pleasure.'

In 1970 we started taking our foreign trips. We took a month-long trip to England, Israel and France. From then on, every two years, we would take a month trip to Europe. We ultimately took a total of seven trips. We then took a one-month trip to Japan and a five-week trip to Australia and New Zealand, as well as a visit to some of the Islands. In the last 20 years alone, we have taken several trips.

One trip, we went to London on business. From London we went to Holland and drove up the coast to Denmark. We then took the car ferry from Denmark to Norway and then took a small boat ride up to the north of Norway and back. We flew to Paris on another trip and spent an entire month traveling around France, Switzerland, Austria, and Northern Italy. We also spent some time seeing the Alps. Two trips were spent in Italy, Spain, Portugal, Yugoslavia, Greece, and back to Israel. We never went to Germany. I could not stand the idea of visiting Germany. We also visited Russia.

On my 80th birthday in 2007, I rented a barge on the Canal du Midi between France and Spain. We invited our four kids to join us in Carcassonne, France. We all met at the train station at Carcassonne, then jumped on the barge and sailed down to Midi over nine days. We had a wonderful time! We then split up and came home.

Shirley and Seymour in Paris 1980s

Shirley and Seymour in Sequoias 1990s

Since then, aside from local trips to Yosemite, etc. we have stayed close to home. During a lot of our trips, particularly in Europe, we collected a lot of treasures which carry many precious memories from our travels.

CHAPTER FOURTEEN: How I Ended Up in Montecito

We had been enjoying our cabin in Big T, and often, we visited every other week. I loved it! Shirley loved it less. One fine day in 1987, I told Shirley "For 30 or 40 thousand dollars, we can enlarge the basement which is vacant and add a lot of room to the house, and I'd like to add a two-car garage outside of my workshop."

Shirley said, "why?"

I said, "I want to move here and move away from LA because it's just too goddamn noisy there!"

Shirley said, "I will not move here; I will stay in LA." She continued "We have a beautiful house on Troon Avenue and if you don't want to live there and you want to stay married to me, we have to figure out something else. But I guarantee you, I will never live one summer here!"

You see, Shirley did not like the heat, and the summers at the cabin were nearly unbearable being that Tujunga Canyon was mostly desert.

Over the next few months, we started traveling along the coast, and eventually went westward and ended up in Montecito. We fell in love with this charming town. We decided to visit often and eventually in 1989 we found a little house there on Veloz Ave. We bought it, said goodbye to Los Angeles and moved to Montecito for good.

We lived there for about three years. The problem though, was that it was much too small for us. We often had Ellen and Tom, Karen and her husband Steve, as well as friends of ours up to visit and I often had to turn my office into a bedroom. Plus, I had absolutely nowhere to work on my own shop projects which I was anxious to do.

Seymour, Bob & Doris (Shirley's Sister & Brother-in-Law), Tom, Ellen, Steve, Shirley & Karen. Early 2000s

So, I started looking for larger places that would work for us and the price was just not right. One day a real estate lady took me up to Knollwood Drive and showed me an empty lot which was over an acre with a lot of trees on it and no house. The place was owned by a speculator who was planning to build a huge house on the lot. We looked at what he was proposing to build and immediately threw his idea out. But we both liked the property. We ended up buying the land and a bridge that gave us access to the property from the road leading up to it. From there we began to design our ideal home.

After a month, I was referred to an architect who looked at the lot, looked at my design, and after a week he recommended that there was a perfect spot to have a second-floor bedroom. We went with his plan, and built a 4,400 square foot, three bedroom, three and a half bath home that included a large office.

I had quotes from three different builders, and I decided that the house should be made from heavier construction than the standard with all the rooms extra insulated. I was told that would cost $130 per square foot. Adjoining the house was a 1,300 square foot, cheaply built, oversized three car garage, plus my new shop, which was about 650 square feet.

I had carefully planned the building of my shop to allow me to make anything I wanted. Everything was built in a little over a year and the total cost came out to $1,500,000 including the land. We moved in May of 1994. I rented a big trailer and dragged it with my truck back to our cabin in Big T, loaded it up with the furniture that we wanted to keep, and brought everything up to Knollwood. In the meantime, we had also moved everything from Veloz to Knollwood as well. Finally, we had enough room for everything! We had a view of the islands and a lot of beautiful oak trees. We enjoyed Knollwood for over 22 very happy years.

Knollwood Dream House Front View

Lush Landscaping Around Knollwood

Seymour in His Office at Knollwood – aka 'The Nerve Center'

In 2012, I came down with a case of macular degeneration. In two years, it got much worse and finally I called Ellen who had moved to a beautiful house about seven miles away from Knollwood. I explained to her that I was no longer able to do bookkeeping. It was hard for me to read but I was still in business and had work that needed to be done that I knew she could do. She said, "Sure pops, I'll be over a couple of times a week." And she was. She would visit a couple times a week, have lunch with us and do my work. This went on for about two years.

CHAPTER FIFTEEN: Seymour's Retirement

A couple of years after Ellen had started helping me with my work at Knollwood, she said, "Dad, you need to start thinking about retiring."

I said, "Retiring? You gotta be crazy!" You see, I was very happy living at Knollwood; it was my paradise. And now, 22 years later my daughter is telling me to move out. I hated the idea. On the other hand, Shirley was more open to it. She had always done all the shopping and the cooking and never complained once about doing it. However, the thought of having someone else take over all that work for her became more and more appealing to her. In retrospect, I realize what I might have done was to add on another room and hire full time live-in help who could have taken over all that work and we would have had the support we needed.

Hindsight is 20/20 however, so in early 2017, we moved into a tiny one-bedroom apartment in Casa Dorinda, which was all that was available at the time. It was way too small, and I hated it. It was very difficult for me to go from having all the space I needed and wanted to this tiny cubicle of a living area. I had to sleep on a cot in the living room while Shirley slept in the bedroom.

I cried like hell to the management to get me a larger apartment. I kept persisting and bothering them, and finally Kendra, the sales manager, came to me and said, "Seymour, in about a month I think I will have a two-bedroom apartment for you!" I was thrilled and soon after, we moved into a large two-bedroom apartment, which was four times larger and very pleasant, in a great location on the campus. I determined that Casa Dorinda was where we should be living after all. And to prove it, I made a large donation, necessary to complete a new construction job on the campus of a new Memory Care Center and updated Personal Care Center. The bottom line is that I put my money where my mouth was.

We still had a problem however of where we were going to put all the stuff we had accumulated during our 68-year marriage. Plus, where in the world would I put my shop and all my equipment and machinery? So, we bought a small house a half mile away from the Casa—which has been known as "the little house". We moved all our stuff that would not fit at the Casa from Knollwood to the little house in September 2017. The shop takes up the two-car garage. One of the bedrooms is the office, another is my painting studio. I have a lot of storage space and room to display my work.

On January 9, 2018, there was an enormous debris flow, which has been called the '100-year flood.' There were huge piles of mud all over the roads in Montecito. Twenty-three people were killed in the flood and we both would have been killed if we had stayed at Knollwood. So basically, we are both living on borrowed time. We have been at the Casa for four years now in complete happiness.

CHAPTER SIXTEEN: The Set Back

Three months after we moved into Knollwood, Karen called Del and said, "The three of us are here at the hospital and my dad is being operated on for a very serious heart condition." Karen told Del it was called a triple bypass operation.

A couple of weeks later, Del called up and Shirley talked to him about how I was doing, and she said, "He's up and around but he's kind of dragging."

Del said, "I wanted to come up and see your new house, but I want to wait until Seymour is completely well so we can spend a lot of time together."

Shirley said, "I'll have Seymour call you when he's up and around."

I heard the story and a few weeks later I called Del up and we talked about him coming up to see us and Shirley said, "We'll have lunch here." This happened a couple of months after the operation. Del was driven up by his son-in-law, Jerry Ivory. Jerry is a stone mason who does fancy work all over town and does very well for himself. He is actually the son-in-law of Del's third wife, Diane.

After I took Del out to see my shop, we sat down and I asked him "What the hell happened at Punch Press Products? I heard that you almost went broke."

Del said "No that's bullshit, but I'll tell you the story. A couple of years ago I hired a crackerjack salesman named Johnson who could get me a lot of work from Sony, and he did. One day, Johnson walked into my office and told me that Sony is going to come out with a new 60-inch TV. Johnson said that the 60-inch was going to sell for about $30,000 and Sony was building an assembly factory in Northern

Mexico. They wanted to have all the parts made in the US and shipped to Mexico for assembly. The bottom line is that I got $4 million in purchase orders from Sony to do this job."

Johnson had explained to Del what was required of him to get the work which was very involved and Del went ahead and invested almost a half a million dollars in the job and started making parts. I told Del that that sounded pretty good to me.

He said, "Yeah, it went great at first, and it looked like I was going to make a mint. Then the shit hit the fan. One day I came to work a little bit late and there was a uniformed guard sitting in front of my office with a big package under his arm. He asked me if I was 'Delmo Charles Molinari' and I told him, "yes, I am."

He says, "Well I have to have your signature that guarantees that you have received all the literature that I have here."

Del said, "I am D.C. Molinari, what you got here?"

The officer handed it over and Del saw that it was in fact a cancellation of all orders from Sony, effective immediately. After Del read everything in the package, he immediately called up Dave Ostrove, his lawyer, and explained to him what was going on. Del said that Ostrove told him to not get excited and he would be over within a half hour, and he was. Dave asked Del to give him the copies of the original purchase orders, and he spent about an hour reading the new information that Del had signed for.

He then looked up at Del and said, "I'm sorry to tell you, but you're screwed, and you lost the job. But they have to pay you for all the work done up until now."

Del said, "I have $4 million worth of purchase orders from them and you're telling me they can cancel everything at will?"

Dave said, “It’s very clear on these purchase orders that there is a cancellation provision that can be exercised, and you’re screwed.”

Del told me within two days he had to lay off 90 people. Del’s total employment went down to 125 people which was exactly the number of people he had started out on his own with the year after we split.

Within a couple of months of all this happening the news came out that Sony and three other companies were coming out with flat screen TVs which had made the older technology obsolete, and that’s why Del had lost the orders.

This was a rough blow for Del, and while he did enjoy several more years of successfully running his company, his business was never quite the same again.

And after my heart surgery, fortunately, I recovered well and continue to enjoy great health to this day.

CHAPTER SEVENTEEN: My Twenty-Five Year History at The Music Academy Of The West

Part A – Renovating the Campus

Shortly after we had moved to Montecito in 1990, I saw an ad in the local paper with Marilyn Horne's picture announcing that she was going to conduct a vocal Masterclass at the Music Academy of the West at Abravanel Hall. We had never heard of the Music Academy of the West, but we had certainly heard of and seen Ms. Marilyn Horne. We attended the Masterclass and were very impressed with Ms. Horne's voice. We were less impressed with the quality of her students.

Marilyn Horne Performing at the MAW

Over the next couple of years, we attended several performances—many vocal, but also instrumental in both Abravanel and Lehmann Halls. I had spent much time snooping around the grounds. Aside from Claeyssens Hall, which was a modern, beautiful building, almost all the other buildings were garbage. They were old, made of wood and their roofs leaked. They stunk when you went inside. They had pianos standing there that had been ruined. There were many beautiful, old landscape features on the campus, but everything desperately needed to be maintained and updated.

After a couple of years, I had not only met Ms. Horne (which was a thrill), but I had met David Kuehn, and Lee Luria and her husband Eli. I kept squawking about how things appeared, and finally to shut me up, Lee asked me to become a member of her Scholarship Committee. This was in 1995.

I was told by a longtime supporter that if we have 100 students and if each paid $10,000, that would give us an income of $1,000,000 per year. Now that statement is true and it stuck in my mind, but in effect it was a bunch of baloney, and it was several years later (after I was a Board member) that I found out what the actual income was. In other words, that was strictly pie in the sky and a convenient target. The Scholarship Committee was formed years before to raise the money for qualified students who did not have the means.

About this time, Lee had hired Scott Reed, who was an intern at UCSB, to become the MAW Dean of Students. He was an instant success. He remained the Dean for two years. After that, he graduated from UCSB.

Marilyn Horne and Scott Reed

Marilyn Horne appeared at a Scholarship Committee meeting and announced that she had just returned from the graduation performance at the Curtis School of Music in Philadelphia, which is a free school for qualified students. She had heard a highly talented young, 23-year-old Israeli Mezzo Soprano. Her name was Rinat Shaham. She insisted that Rini be brought to MAW the next year.

Lee looked at me and said, "Seymour, here's your chance to put your money where your mouth is."

I said, "OK, I'll do it."

I did not know that Rini was going to come for three years in a row, but it was the best money I ever spent—and by the third year Rini practically owned the vocal program. Everyone loved her, so did we. She brought heavenly music into our lives and became one of our dear friends.

Rinat Shaham Singing at a Vocal Recital at the MAW

At the end of Rini's third year, which was in 1997, the vocal department was going to have a performance of a great Italian opera, in which Rini played the lead role. The singers had rehearsed for six weeks. In the seventh week, the performance was going to be held at the Lobero Theatre. I made a reservation at an Italian restaurant that has a separate room that could hold about 20 people. I booked the entire room. The Friday before the performance, I drove to the Academy and picked up the five performers who were playing the leading roles, including Rini. We drove to the restaurant and had drinks and hors d' oeuvres before dinner. Rini started humming her role. The other singers at our table did the same. The rest of the room was filled up with other vocal supporters and their fellows. Suddenly Rini stood up and in full voice started singing her role. Everyone else did the same in answer to her. Several of the waiters, Italians of course, also started singing. The entire restaurant, aside from our room was also singing with a lot of vocalists. Shirley and I were listening to this, and I thought I had died and gone to heaven. It is a moment I will never forget.

Over the next twenty years or so, we sponsored 36 students.

The Scholarship Committee took on the additional function of entertaining the students, and later the faculty and supporters on each July 4th holiday, culminating in 1999 with a big celebration at the Santa Barbara Zoo.

In 1998, a group of supporters and members of the Board of MAW formed a committee called the Renaissance Committee for the purpose of rebuilding the campus. They spent a lot of their own money with lawyers, architects, and contractors to develop a very ambitious plan. In the year 2000, I was elected to the Board of the MAW. During this time, we spent most of every monthly meeting analyzing every detail of the Renaissance plan. It turned out that it was totally unrealistic and unworkable, so we threw it out.

Also, in the year 2000, David Kuehn proposed that the MAW be a free school for qualified-only students. The Board brought up the issue that this would require another $500,000 in donations and didn't think that this would be possible. We realized that we had a big problem and we needed help. It was a big struggle.

By this time, around 2004, Scott Reed had been working at the San Francisco Opera in their development department for two years. Lee stayed in touch with Scott during this time. She went on a campaign to hire Scott back to the MAW as the head fundraiser, which we did.

That same year, I was invited to a meeting at David Kuehn's office. He had received a quotation for the reconstruction of one of the worst wooden buildings—a rehearsal building—the name of which ultimately became the Lehrer Studio Building. The price was $800,000 and he asked me to pay for it. I told him I'd have to think about it.

During that week, we had another meeting at which a new Board member by the name of John Burgee was in attendance. John had been a partner in Johnson and Burgee Architects of New York, one of the major architectural firms in the U.S. There was nothing about the building business that he did not know. He reviewed the proposal, we looked carefully at the building and within a week he came back and said that rebuilding the structure according to the proposal was an absolute waste of money. Shortly thereafter he brought in Schipper Construction. We all had a lot of discussion about the job, and they came back with a very detailed proposal, which John recommended. The price was double the previous price. The Board approved the new proposal, with the understanding that I was going to pay the $800,000 and they would pay the rest. Ultimately the price of the building wound up at $2,300,000. $100,000 each was donated for twelve of the rooms. Four of the larger rooms were paid for by larger donations.

While the work was being done, a lady came to me and said, "Seymour, I want you to know something. Pierre Claeyssens bought twelve new Steinways for his hall at the completion of his building. Are you going to do the same?" I went to Scott and told him the story and he immediately called Steinway and Sons and the manager came out for a meeting with catalogs and literature. He inspected all our pianos and recommended that he take each Steinway in trade for 13 new ones that we wanted to buy. He also recommended that many of the other ones be hauled away for scrap. The eight that he took in trade were rebuilt as good as new, and we purchased ten new additional student pianos.

In 2005, David Kuehn resigned as President and was replaced by Nancy Bell Coe, who remained the President for six years. At the conclusion of which, Lee Luria and I went on a campaign to elect Scott as the new President of the Music Academy. That took place around 2010.

While all this was going on, John Burgee was quietly studying the construction of Abravanel Hall. I was personally involved in all the campus construction. In over a year's period he told me he thought it could be rebuilt and save some money by using the original structure. He came up with an approximate price for turning the building into a proper performance hall. This took about two years' time. The price turned out to be $8,000,000. Scott did a miraculous sales job to Stephen Hahn, who promised to pay $4,000,000 plus he would contribute $1,500,000 for an endowment fund to be used for the maintenance of the building into perpetuity. To say the least, Scott was very busy. The beautiful building now called Hahn Hall turned out to be an architectural wonder. It is an amazing 300 seat concert hall with acoustical tiles that John Burgee designed, making perfect acoustical quality, along with all the decoration.

In 2010, I convinced Tom Orlando, my son-in-law, to become a member of the Board and he agreed, not knowing what he would ultimately be involved in.

Seymour, Scott, Shirley, Tom & Ellen at a MAW Fundraiser

Also, in 2010, the Lurias got very serious in the design of the proposed Luria student center—a major undertaking which took two years. It involved the rebuilding of the northern portion of the main building, going down to its foundation. The construction consisted of a masterclass venue, seven individual practice rooms, two chamber ensemble rooms, a teaching studio, an expanded music library and listening areas, a firesafe room for archives, a catering kitchen, an outdoor dining pavilion and new restrooms. In addition, there is a large space for performances and meetings called Weinman Hall, dedicated to supporter Robert Weinman. It turned out to be a magnificent addition.

Around 2014 it was proposed that the remaining ramshackle buildings be torn down and replaced with a new proposal, which ultimately became Hind Hall, named after the Hind family of San Luis Obispo. With the Hind major construction completed in 2016, basically everything had been done except the reconstruction of the main building which began around 2016. By that time, Tom had taken over for me in my duties on the Construction Committee work. The reconstruction of the main building turned into an unbelievable nightmare. I'm certain it shortened Tom's life and is almost impossible for me to retell.

In September 2017 Tom and I made an appointment with Scott in his new office in the main building. Scott ushered us into his office. I was sitting on the couch bouncing Scott's baby, Ruby, on my knee. He asked us what he could do for us.

I said, "Scott, I'm here to resign from the Board. My membership is of no value to you. I can't see across the room to who is talking, and I cannot read, so you should replace me with somebody that could do you some good."

Scott was silent. But then we began talking about other things and about fifteen minutes later, Scott sat down opposite us and told us that he would not accept my resignation. I was shocked. I asked him why not. He said, "Simply because you're the only one here that knows the whole story."

Well, that knocked me for a loop. He also told me that I have Tom who can read, and who communicates with me, so I will know what's going on, and I will only have four or five more meetings in the years 2018 and 2019. My 19-year Board term was up at the end of 2019.

I told Scott that I wanted to write a brief history of what had happened during the previous 25 years, and he said, "please do." So, here it is.

In this first part I am retelling all that went into reviving and reconstructing the campus and its buildings up to its present state. In addition, a great deal went also into the development of the artistic programs of MAW and transforming it into the world class academy that it is today.

Seymour and Karen Lehrer at a MAW Fundraiser
Karen, my daughter, provided much moral support during my time at the MAW

Rinat Shaham with Steve Sherwin, Shirley, and Karen

Cutting the Ribbon on the new Lehrer Studios From left: Tom and Ellen Orlando, Shirley, Seymour, Karen Lehrer, Steve Sherwin

Shirley and Seymour in Front of the Newly Renovated and Dedicated Lehrer Studios

Early Rendition of the Lehrer Studios

Lehrer Studios as it is today

Part B - The Evolution of the Music Program

The re-building and construction of the buildings of MAW certainly put a beautiful new face on the campus which encouraged donor support and impressed strangers. However, it was NOT what truly made the Music Academy that we now have today. The significant transformation and growth of the artistic quality of MAW's programs all started in the year 2,000 with the opinion of David Kuehn that we needed to do more in the way of auditioning new candidates. This started a practice that continued to be improved over the years and ultimately led to many perfections in the work. Number one, we got better music teachers who participated in the selections. This led to a vast increase in the number of applicants, which at times numbered up to 2,000 in a year.

Of the most difficult selections is the vocal program. At times we've had up to 800 vocal applicants for a total of 23 positions, or 3% admissions. Of course, I believe that almost immediately, many of the singers were cast out without them even having had a chance to audition.

The pianists also had difficulties and about 10% of the applicants succeeded. The string players had an equal challenge. The competition was intense and at the end of each year's selection process, only 140 musicians out of 2,000 applicants earned acceptance into the MAW program.

The result of this is that by June 2006 our orchestra conductor was able to have the best of the best, and ten days after his musicians arrived for the summer program, he conducted his first concert at the Covenant Church. The entire audience was shocked that in only ten days he had put together a very professional sounding group.

Gisele Ben-Dor, who was the conductor of the Santa Barbara Symphony, was sitting next to me and when the program was over, she looked at me and said, "This is absolutely amazing!" She was especially impressed because the conductor had announced that only a week and half before, most of the players had to sight read the music as they had never seen it.

When Scott Reed came back to us in 2004, he made a point of becoming close to some of the Board members that had far reaching knowledge and ideas about the future of the MAW. They became members of his kitchen cabinet. When he became the President, he was able to bring in new people with wide ranging experience, in particular, Ana Papakhian, who was the assistant to Marilyn Horne. These visionary people laid the basis of what needed to be done.

Initially, one of the most important ways we needed to support our Fellows was to set up new living quarters for them. For years we had discussions about moving away from Cate School, the Carpinteria school that housed our Fellows. The facility had many disadvantages. Among them was the fact that it was a long distance from the campus. The other big issue was that it was a high school, not a college. We considered every possible housing facility in Santa Barbara. No place would be appropriate except Westmont, which is a ten-minute bus ride from our campus. The problem was that Westmont is a very strict Christian college with very precise religious constraints, which unfortunately our adult Fellows could simply not adhere to. After three years of careful coaxing and expert negotiating by Scott, we were able to work out an arrangement that worked for both organizations – a real miracle. Lee Luria's daughter, Candace (Candy) whom Scott had hired to head the campaign, was also immensely helpful.

Another reason we really wanted to work out a deal with Westmont is that it has a well-developed music department with 20 Steinways, many practice rooms, several recording facilities, and a beautifully maintained campus. Happily, Westmont has continued to be available to our Fellows for the summer festivals since 2017.

Before the efforts for Westmont began, Scott and his kitchen cabinet came up with the idea of forming a partnership with a noteworthy orchestra. After a lot of consideration, they decided on the New York Philharmonic. Some people thought he was nuts. But the bottom line was that around 2014, the NY Phil came to Santa Barbara, and with 60 of our Fellows performed at the SB City College football stadium, in a temporary facility overlooking the ocean with an audience of 7,000. It was a smash hit beyond belief and counts as another one of my significant memories being involved with MAW.

In addition, the contract with the NY Phil provided that ten or twelve of our Fellows in turn would perform in New York with the orchestra and get scale wages. Can you imagine what it would mean to a young musician working under Alan Gilbert?

Next was the two-year affiliation set up with the LSO, London Symphony Orchestra. They came to Santa Barbara in 2017 and performed in two separate summer festivals.

All these things in my estimation have raised the MAW to its present vaulted position in the music world, a position that only continues to grow into the future with the excellent guidance of the current board.

In September of 2019, my 20-year presence on the board of the MAW came to a close. Scott and the board threw a farewell luncheon in Weinman Hall in my honor. They sat me in front of a large screen, the reason for which I did not know. Suddenly Rinat Shaham (Rini) appeared on the screen standing in front of the Brussels Opera House, where she was performing at the time. Since her years at the MAW, she had gone on to have a tremendous music career. She was being interviewed about the years she had spent with us training at the MAW when we had sponsored her. At one point she became teary eyed, and she was speaking so from the heart that I broke down too. Soon everyone in the audience was crying.

Seymour and Shirley at his MAW Awards Dinner

Seymour Accepting His Steinway Piano Key Award from Scott Reed / MAW

Over the years, we have sponsored 35 very talented musicians, and recently, the Lehrer Foundation made yet another sizable contribution to the Vocal Program at the MAW. Many of these students go on to highly successful careers in music. The achievement of this as well as being a part of the revival and growth of the Music Academy, I count as one of my greatest blessings this life.

CHAPTER EIGHTEEN: Reflections...

As I have reflected on my history while writing this book, I feel extremely fortunatc that I have been able to make a wonderful life out of my passion for metal working. Additionally, I have been able to apply my skills creatively in a longtime hobby of making metal sculptures. In essence, I've been able to live a full well-rounded life both in business and in creative craftsmanship.

However, though I had a successful multi decade career as a metal worker, it has saddened me to an extent that in my retirement here in Montecito I have never met anyone who knows anything about my industry. Back in the day, I was used to 'talking shop' with just about everyone I knew. They were all in different aspects of my industry. Every once in a while, I'll meet someone who is a mechanical engineer who knows something about metal fabrication, but this happens very seldom.

My hope is that by sharing my story, I can inspire the reader to a greater appreciation of how much the metal working industry is to the comforts and lifestyle we all enjoy and take for granted today.

Seymour Lehrer at the book signing for
'Poems and Stories From the Lucky Life of Seymour Lehrer'

"... and to my brother-in-law Seymour, who wanted to be remembered in my will—hi there, Seymour!"

Compensation

By Edgar Guest

I'd like to think when life is done
That I had filled a needed post.
That here and there I'd paid my fare
With more than idle talk and boast;
That I had taken gifts divine.
The breath of life and manhood fine,
And tried to use them now and then
In service for my fellow men.

I'd hate to think when life is through
That I had lived my round of years
A useless kind, that leaves behind
No record in this vale of tears;
That I had wasted all my days
By treading only selfish ways,
And that this world would be the same
If it had never known my name.

I'd like to think that here and there,
When I am gone, there shall remain
A happier spot that might have not
Existed had I toiled for gain;
That someone's cheery voice and smile
Shall prove that I had been worthwhile;
That I had paid with something fine
My debt for a life divine.

Acknowledgements

I would like to thank Ellen Orlando and Karen Lehrer, my daughters, Tom Orlando and Steve Sherwin, my sons-in-law, Doris Schaffer, my sister-in-law, and my editor, Samantha Bonavia for their individual contributions to this book.

Made in the USA
Monee, IL
04 November 2026

67380768R00069